Membrane Separations Technology

Membrane Separations Technology

Single-Stage, Multistage, and Differential Permeation

E. J. Hoffman

G|P / P|⍦ Gulf Professional Publishing

An Imprint of Elsevier Science

Amsterdam Boston London New York Oxford Paris
San Diego San Francisco Singapore Sydney Tokyo

Library of Congress Cataloging-in-Publication Data

Hoffman, E. J. (Edward Jack)
    Membrane separations technology: single-stage, multistage, and differential permeation / E.J. Hoffman.
        p. cm.
    Includes bibliographical references and index.
    ISBN 0-7506-7710-4 (acid-free paper)
    1. Membrane separation. I. Title.

TP248.25.M46H64 2003
600'.28424—dc21   2002192890

British Library Cataloguing-in-Publication Data
A catalogue record for this book is available from the British Library.

The publisher offers special discounts on bulk orders of this book.
For information, please contact:

Manager of Special Sales
Elsevier Science
200 Wheeler Road
Burlington, MA 01803
Tel: 781-313-4700
Fax: 781-313-4882

For information on all Gulf Professional Publishing publications available, contact our World Wide Web home page at: http://www.gulfpp.com

10 9 8 7 6 5 4 3 2 1

Printed in the United States of America

Contents

Preface

Techniques used in steady-state flash vaporization calculations and multistage distillation calculations can be utilized to show that membrane separations are enhanced by the use of cascade or multistage operations. This is of importance particularly in the use of membrane materials showing low selectivity between the components to be separated.

The primary domain of interest here is gaseous separations rather than liquid separations, nor are vapor-liquid separations necessarily considered, other than that there is a similarity in the equation forms used, as is emphasized in Chapter 3. Nor are gas-solid or liquid-solid separations considered, except in passing.

It is first assumed that perfect mixing can be attained for single-stage and multistage or cascade membrane operations and that a condition of constant molal (or molar) overflow or underflow can be approximated during multistage operations.

The calculations, in this respect, become similar to those employed for single-stage flash vaporization and multistage distillation with reflux and reboil, or for absorption or stripping. All the latter utilize the concept of an equilibrium stage. It should be emphasized, however, that the adaptations are constituted to apply to the nonequilibrium rate phenomena associated with membrane permeation. The calculations are similar in form but not in content. For one thing, the permeate flow rate per unit of membrane area (that is, the permeate flux) becomes part of the distribution coefficients or K-values for each component. An extra element of trial and error is therefore introduced.

Of significance is that multistage membrane separations for binary systems can be treated graphically, in a fashion similar to the classical McCabe-Thiele method for binary distillation. This is developed and illustrated in Chapter 4 and affords a convenient means for evaluating separation possibilities, in determining the effect of permeability, reflux or recycle ratios, the number of stages, and feedstream location, as further detailed in Chapter 4.

Following this, more rigorous derivations for differential permeation are pursued. That is, permeation is viewed as a continuum rather than an operation in discrete stages or cells. In the first case to be considered, there is continuous point permeate withdrawal; in another case, the more general case of permeate flow is analyzed, both concurrently and countercurrently.

Finally and foremost, simplifications are provided, especially for operations with recycle or reflux of the product streams to induce a sharper separation. For countercurrent differential permeation, the correspondence becomes similar to distillation and absorption or stripping in a packed or wetted-wall column, although the representation and the calculations can become much more complicated, unless simplifications are made.

The objective, therefore, is to subject the processes of permeation to a rigorous and systematic analysis, with appropriate reductions or simplifications, which will permit the process design of membrane units. That is, given the component permeability and the degree of separation specified, the number of stages or the flow juxtapositions, and the membrane areas can be determined. Or, given the latter, the degree of separation can be found—a much more straightforward route. These are the criteria normally encountered in process design calculations, and it is the purpose here to systematize membrane calculations so that membrane separations can routinely be accommodated as a chemical engineering unit operation. Membrane reactors are treated as a special case.

The appendices contain spreadsheet calculations that are systematized, corresponding to each chapter of the text. The successive relationships involved in each case or embodiment are provided in spreadsheet notation, so that each methodology can be readily entered into and conducted on a personal computer using the requisite software (e.g., Excel or Lotus 1-2-3). In this way, the calculations can be adapted to whatever membrane characteristics and operating conditions are to be specified, both for single-stage and multistage separations with recycle or reflux, and with the entirety considered as a continuum. The last-mentioned involves either concurrent or countercurrent behavior, without or with recycle or reflux. Moreover, the means are furnished, not only for determining the degree of separation, but the membrane area requirements.

1

Introduction

Membrane separations are an accepted means for separating non-condensable gases; that is, gases that ordinarily condense only under low-temperature or cryogenic conditions. The technology might be in wider use if (1) better and more selective membrane materials were available and (2) the necessary mathematical representations and calculations were better spelled-out for the separations attainable. In fact, the one sometimes depends on the other. Of particular interest are ways in which separations could be enhanced using relatively nonselective membranes.

A special case is pervaporation, in which the feed material is a liquid but the permeate is a gas. That is, the temperature and pressure of the permeate are such that the permeated components exist in the gaseous phase. Conceivably, however, the feedstream could be a gas but the permeate conditions and compositions are such that the components constitute a liquid phase. For the particular purposes here, however, all streams are in a gaseous state.

The general subject has been explored in a number of past reviews[1-8] and, for instance, is the main concern of the *Journal of Membrane Science*. The subject has also been of interest to the Gas Research Institute, which has held workshops on the subject.[3,4] The Gas Research Institute, in fact, jointly sponsored a project with the Dow Corning Corporation and others aimed at correlating and predicting the permeability behavior of membranes from the chemical structure.[9,10]

The American Institute of Chemical Engineers has maintained an active interest through its Symposium Series.[11-13]

Membrane Processes in Separation and Purification,[14] published in 1993, contains chapters on pervaporation, facilitated transport membrane processes, membrane gas absorption processes, hollow fiber contactors, membrane reactors, and the preparation and application of inorganic membranes. In addition to an introductory chapter by the editors, *Polymeric*

Gas Separation Membranes,[15] published in 1994, has chapters on the following subject areas: the diffusion of gases in polymers, the relationship between polymer structure and transport properties for aromatic materials, the relationship between polymer structure and transport properties for high free-volume materials, the formation of membranes specifically for gas separations, a discussion of facilitated and active transport, nonhomogeneous and moving membranes, membranes for separating organic vapors from gas streams, gas separation practices in Japan, further commercial and practical aspects of gas separation membranes, and a comparison of membrane separations with other gas separation technologies. Neither of these volumes details the process-type calculations involved for determining the degree of separation.

As to a comparison of membrane gas separation technologies with such methods as pressure swing absorption (PSA) and low-temperature or cryogenic separations, the last-cited chapter in *Polymeric Gas Separation Membranes* must remain somewhat inconclusive, given the wide range of variables, parameters, and applications.[16] Moreover, for the most part, the separations compared were confined to air and hydrogen-containing systems.

More recent publications include *Membrane Separations Technology: Principles and Applications,*[17] published in 1995 and edited by Richard D. Noble and J. Douglas Way, who had coedited an earlier volume, *Liquid Membranes: Theory and Applications.*[18] The state of the technology is kept track of by the Business Communications Company, for example, in *Membrane and Separation Technology Industry Review,*[19] published in 1998. For continuing developments, consult *Books in Print* and WorldCat, a service in conjunction with OCLC (Online Computer Library Center). Additionally, there is, of course, the Internet.

A review of developments and directions, as of 1991, was published under the auspices of the U.S. Department of Energy.[20] Apart from the more-inclusive technical information sources furnished by a computer search of Chemical Abstracts, the National Technical Information Service keeps a running account of government-sponsored research, including bibliographic updates, listed by title and author(s), with other particulars, and with an abstract (NTIS Bibliographic Database, available on compact disc, updated periodically). The entries for membrane-related research in general number into the thousands, much of it biomedical, and the entries for membrane gas separation constitute only a relatively small part.

In particular, the entry for Membrane Gas Separation (as per citations from the NTIS Bibliographic Database) is abstracted with a lead-off qualifier, to the effect that, "The bibliography contains citations concerning the research and development techniques involving the use of plastic and metal or metallic membranes." A specific example of an additional statement of scope (June 1993) is, "Included are such topics as recent advances in membrane science and technology, gas separations using composite hollow fiber membranes, optimal cascade theory for the separation of mixtures on semipermeable membranes and gas separation by a continuous membrane column." Another example (August 1993) is, "Citations review isotope separation, osmotic techniques, reverse osmosis, and preparation of membranes for specific separation processes. The permeability of polymer membranes is discussed in terms of physical properties as well as molecular structure." And, in closing, "The selectivity of polymeric films for a variety of gases is also included." A subject or terms index and title list are included. As the examples will indicate, the coverage is extensive.

Representative samples from the NTIS data file, which give an indication of some of the directions in which membrane research is headed, include the references about coal-derived gases and liquids and their further separation or conversion,[21–27] in particular high-temperature ceramic membranes and the use of membranes as catalytic reactors. Similar remarks could be made for metallic membranes; for example, the diffusion of hydrogen through metals is a subject of long standing. Membrane separation processes in the petrochemical industry, for instance, are reviewed in *Membrane Separation Processes in the Petrochemical Industry*.[5] With regard to the separation of liquids, some recent developments are presented by Cabasso et al.,[28] and for the separation of solids and liquids by hyperfiltration, by Leeper and Tsao.[29] The commercial implementation largely remains to be seen.

Of notable mention, the *Membrane Handbook* was published in 1992.[30] Another work of interest is *Membrane Separations Technology: Principles and Applications*, edited by R. D. Noble and S. A. Stern, a volume in Elsevier's Membrane Science and Technology Series.[31] Also, appropriate entries can be found in the *Encyclopedia of Chemical Processing and Design* (Volume 27)[32] and the *Kirk-Othmer Encyclopedia of Chemical Technology* (Volume 15).[33] Evidently, the subject has not yet reached the status of a chemical engineering unit operation, since the necessary process-type calculations are as yet ill-defined or undefined. For instance, consult the section on membrane separations in the seventh edition of *Perry's Chemical Engineers' Handbook*.[34] Notwithstanding all of this, the principal item of interest here is not in membrane materials or membrane cells per se, nor in the usual considerations of membrane science and technology, but in the derivations and process-type calculations involved in predicting the degree of separation that can be attained. It is a matter more complicated than ordinarily thought or expected and a fitting continuation of the unit operations concept as embodied in chemical and process engineering.

For the record, however, a few background preliminaries about membranes will first be introduced.

1.1 MEMBRANE MATERIALS

A considerable array of membrane materials exist for various gaseous separations, some more effective than others.[10] That is, some are more permeable and more selective than others. It depends on the system to be separated, however. In other words, materials are not yet available for the full array of gaseous mixtures encountered. A partial listing is presented in Tables 1.1 through 1.3, giving properties and selectivity or relative permeability of components.[4,9,10,30,35]

Table 1.1 Relative Permeability for Cellulose Acetate Membranes

Gas	Relative Permeability
$H_2O(g)$ (considered fast)	100
H_2	12
He	
H_2S	10
CO_2	6
O_2	1.0
Ar	
CO	0.3
CH_4	0.2
N_2	0.18
C_2H_6 (considered slow)	0.10

Source: W. H. Mazur and M. C. Chan.[10,35]

Table 1.2 Permeability to Oxygen

Polymer	Permeability (in $10^{-9} \times cm^3/sec\text{-}cm^2\text{-}cm\ Hg/cm$)
Dimethyl Silicone	50
Polybutadiene	13
Polyethylene	0.1
Nylon	0.004
Teflon	0.0004

Source: Gas Research Institute.[3,10]

Table 1.3 Selectivity for Dimethyl Silicone Polymer

Gases	Selectivity $(\alpha = P_i/P_j)$
O_2/N_2	2.0
CO_2/CH_4	3.4
CO_2/H_2	4.9
CO_2/CO	9.0
H_2S/CO	28.0

Source: Gas Research Institute.[3,10]

Table 1.4 Membrane Separations: State of the Technology

Known Separations	To Be Determined
H_2/C_1+	H_2/CO_2
H_2/CO	H_2S/CO_2
He/C_1	NH_3/H_2
$H_2O(g)/C_1+$	NH_3/C_1+
H_2S/C_1+	NH_3/N_2
CO_2/C_1+	SO_2/C_1+
CO_2/N_2	SO_2/CO_2
CO_2/CO	NO_2/C_1+
NO_2/CO	C_1/C_2
NO_2/N_2	N_2/C_1
CO_2/air	Ar/air
	Organic vapors

Note: C_1+ represents methane and heavier hydrocarbons.
Source: W. J. Schell.[10,36]

The oxygen/nitrogen membrane separation for air, perhaps the most obvious, has also been one of the most studied and is sort of a baseline reference. The sharp separation between nitrogen and oxygen on a commercial scale remains in the domain of cryogenics, although membrane separations have been used successfully when only a relatively minor increase in the oxygen content of air is sought, as in portable oxygen concentrators for home use.

The separation of refinery gases is also an item of interest; for instance, gas streams containing hydrogen. In the main, membrane methods pertain to the separation of noncondensable gases; that is, to gases not readily liquifiable except by low-temperature or cryogenic means.

In Table 1.4, the interim state of technology is acknowledged as feasible for various binary separations; future needs are also listed.[4,9,10,28,36] Some of the commercial technologies and suppliers are reviewed and listed.[10,30,35]

Formerly, membrane materials consisted mainly of barrier types, sometimes called *permeable* or *semipermeable*, in which the gases flowed into and through the pores and interstices, which were of molecular dimensions; for example, measured in angstroms. There is the use of materials similar to molecular-sieve adsorbents, for example. For single-phase liquid systems or solutions, the processes may be referred to by the terms *dialysis* and *osmosis*, whereas for gas-liquid, gas-solid, or liquid-solid separations, the terms *micro-* and *ultrafiltration* are more appropriate.

(Some recent developments in membrane processes for the separation of organic liquids are presented in Cabasso et al.,[28] as previously noted; the same may be said for the use of hyperfiltration as applied in ethanol recovery.[29] The latter subject is of relevance also in the processing of nonpasteurized beers; that is, in the separation of spent yeasts after fermentation.)

The more modern embodiment for the membrane separation of gases is the diffusion-type mechanism, whereby the gases actually dissolve in the material and pass through by molecular diffusion. Another embodiment is the facilitated transport membrane, which acts as an absorber on the high-pressure side and as an absorbent regenerator on the low-pressure side. Liquid membranes have also been investigated. Metal or metallic membranes are under study as well, as are ceramics, whereas the usual materials are polymeric in nature. Metallic and ceramic membranes can be used at higher temperatures and may also serve as membrane reactors. Schematic representations of membrane materials are provided in Figure 1.1.

More exotica about gaseous separations are provided by the Gas Research Institute.[3,4] Furthermore, as previously noted, membranes afford the possibility of catalysis.[12,25,26,27]

As also mentioned, a study into the structure-permeability relationships for silicone membranes was jointly sponsored by the Gas Research Institute and the Dow Corning Corporation.[9] An attempt was made toward correlating, understanding, and predicting the permeability behavior of silicone polymers from their chemical structure. This behavior was in terms of the permeability and selectivity to various of the common gases and their separation. The ultimate objective was to systematize and generalize this behavior so that it could be applied to other kinds of membrane materials and to other gases and gaseous mixtures.

1.2 MEMBRANE CELLS

The simplest form of a membrane cell or module is illustrated in Figure 1.2. Called a *plate and frame*, its dimensions are linear in the plane. Commercial applications, however, in the main, utilize tubular constructions.

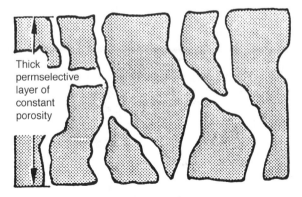

Thick permselective layer of constant porosity

(a) Isotropic membrane

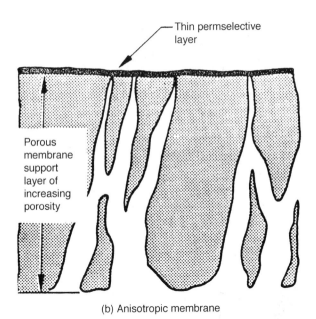

Thin permselective layer

Porous membrane support layer of increasing porosity

(b) Anisotropic membrane

Figure 1.1 Schematic sections of membrane materials. (Source: Leeper et al.[7])

The two main commercial embodiments for membrane cells are illustrated in Figures 1.3 and 1.4. Both are remindful of heat exchanger practices; that is, tube-and-tube and shell-and-tube heat exchangers.

The spiral-wound fabrication shown in Figure 1.3 is analogous to a single tube-in-tube or double-pipe heat exchanger.[37] The hollow-fiber concept lends itself to the shell-and-tube type of heat exchanger. The hollow fibers are mounted as a tube bundle inside a shell, as indicated in Figure 1.4. A few applications are indicated in Figure 1.5.

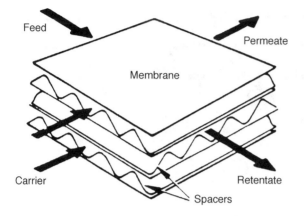

Figure 1.2 Plate-and-frame module. (Source: Leeper et al.[7])

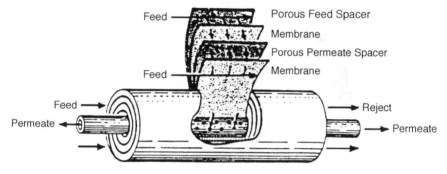

Figure 1.3 Spiral-wound membrane. (Source: Bravo et al.[6])

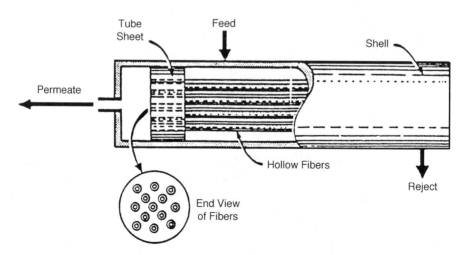

Figure 1.4 Hollow-fiber membranes. (Source: Bravo et al.[6] and Leeper et al.[7])

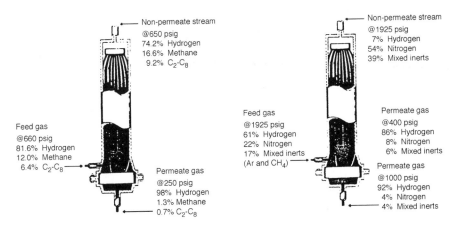

Figure 1.5 Applications of membrane systems. (Source: Prism® Gas Separation Systems, Monsanto Chemical Co., St. Louis, MO; Air Products and Chemicals, Inc., Allentown, PA.) Gaseous components and systems include argon, helium, hydrogen, carbon monoxide/syngas, nitrogen, oxygen, CO_2-removal, H_2S-removal, dehydration. Capacity for nitrogen recovery is up to 35,000 SCF/hr for 97% purity, and about one-tenth of this for 99+% purity.

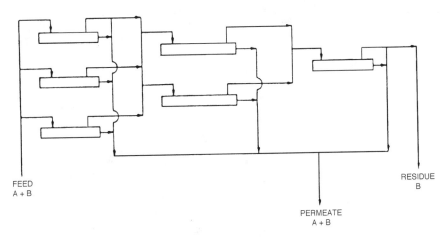

Figure 1.6 Taper configuration. (Source: Hoffman[9] and Hoffman, Venkataraman, and Cox.[10])

1.3 THE ENHANCEMENT OF SEPARATION

With a membrane showing high selectivity between the gases to be separated, a single-stage operation suffices. For membranes of lower selectivity, more involved juxtapositions become necessary. Examples are shown in Figures 1.6 and 1.7 for multistage taper and cascade arrangements.[10]

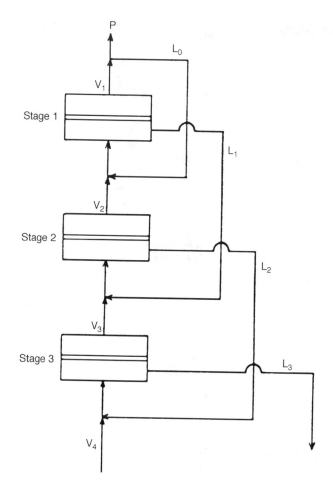

Figure 1.7 Cascade. (Source: Hoffman[9] and Hoffman et al.[10])

The taper configuration of Figure 1.6 does not produce a sharp separation. In this case, only the less-permeable component tends to be recovered in the pure form as the reject. The permeate product is a mixture, although the proportions differ from the feed. The effect is similar to the concept of stripping the more permeable component from the reject phase.

The taper configuration can be changed so that the more permeable component is concentrated in the permeate, whereas the reject product is a mixture. This corresponds to rectification or absorption, in which the more permeable component is concentrated in the permeate phase and the less permeable component is absorbed from the permeate phase. The analogy is to distillation practices.

It may be noted that the cascade arrangement of Figure 1.7, if suitably disentangled, corresponds to a multistage operation as encountered in absorption, stripping, and distillation practices.

Differential Permeation

Of further interest and concern is the operation of a membrane cell as a continuum. Such a view may be referred to as *differential permeation*. The permeate may be withdrawn at points along the membrane, as illustrated in Figure 1.8. Or the cell may be operated in concurrent flow, as shown in Figure 1.9, or countercurrent flow, as shown in Figure 1.10. There is the possibility, even, of producing two permeate products if two different membrane materials are employed separately in the same unit or module. This is indicated in Figure 1.11.

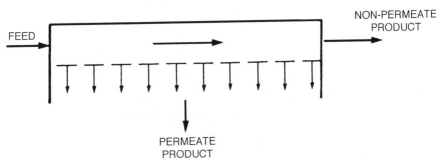

Figure 1.8 Point withdrawal of permeate. (Source: Bravo et al.[6])

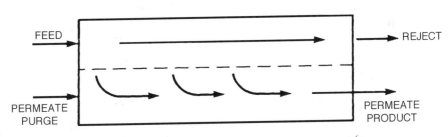

Figure 1.9 Concurrent flow permeation. (Source: Bravo et al.[6])

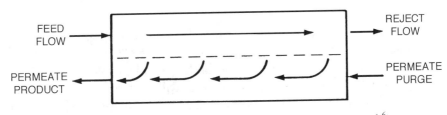

Figure 1.10 Countercurrent flow permeation. (Source: Bravo et al.[6])

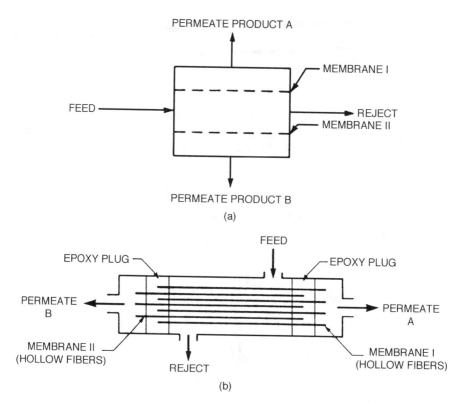

PERMEATE PRODUCT A

MEMBRANE I

FEED

REJECT
MEMBRANE II

PERMEATE PRODUCT B

(a)

FEED

EPOXY PLUG

EPOXY PLUG

PERMEATE
B

PERMEATE
A

MEMBRANE II
(HOLLOW FIBERS)

REJECT

MEMBRANE I
(HOLLOW FIBERS)

(b)

Figure 1.11 Asymmetric permeator configuration. (Source: Bravo et al.[6])

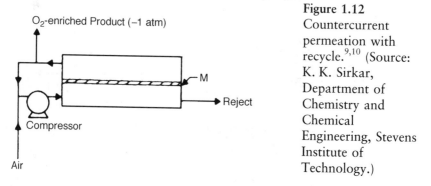

O_2-enriched Product (−1 atm)

M

Reject

Compressor

Air

Figure 1.12
Countercurrent
permeation with
recycle.[9,10] (Source:
K. K. Sirkar,
Department of
Chemistry and
Chemical
Engineering, Stevens
Institute of
Technology.)

Another possibility is the use of recycle in a single-stage cell operating in countercurrent flow, as shown in Figure 1.12.[3,10] More complicated arrangements were shown, for example, at the second GRI workshop.[4] There is the potential here for sharp separations, as will be derived and explained.

Whereas in single-stage or multistage embodiments perfect mixing may be assumed, the use of concurrent or countercurrent flow can also be

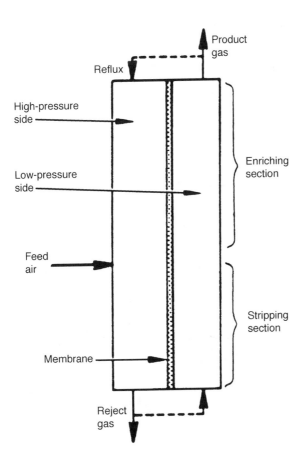

Figure 1.13
Continuous
membrane column
with reflux from both
product streams.
(Source: Leeper et al.[7])

assumed in a context corresponding to an absorber, stripper, or distillation column. This is the case with the system in Figure 1.12. More complicated arrangements may be made, as shown in Figures 1.13 and 1.14. There is the use of reflux or recycle to enhance multistage separation, as indicated in Figure 1.15, which corresponds to the practices of distillation.

A difficulty with whatever the juxtaposition or arrangement may be is the mathematical means for representation and calculation. We are predominantly concerned with this, the necessary derivations, and their simplifications. Of prime importance is the separation that can be achieved. Also of interest is the sizing of the membrane area necessary.

It may be added that the reject or retentate phase for a membrane cell forms a continuum with the feed, assuming perfect mixing. Moreover, this feed-reject phase is commonly pictured schematically as the "upper phase" and the permeate as the "lower phase," albeit both phases are

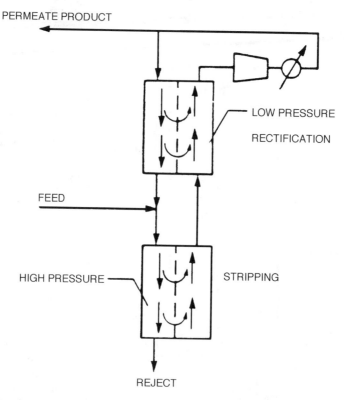

PERMEATE PRODUCT

LOW PRESSURE

RECTIFICATION

FEED

HIGH PRESSURE

STRIPPING

REJECT

Figure 1.14 Continuous membrane column with reflux from the permeate product. (Source: Bravo et al.[6])

gaseous. As matters proceed, we adopt the opposite representation, whereby the feed-reject phase is pictured as the lower phase and the permeate is the upper phase. This is to make the representation more closely analogous to that for vapor-liquid separations and distillation calculations. The derived similarity to the representation of vapor-liquid phase behavior is, in fact, the keystone to systematizing membrane separation calculations.

1.4 SUBQUALITY NATURAL GAS

A potentially large market for membrane applications is the upgrading of subquality natural gas.[9,10,38] Subquality natural gas contains significant concentrations of nonhydrocarbons, which must be partially or totally

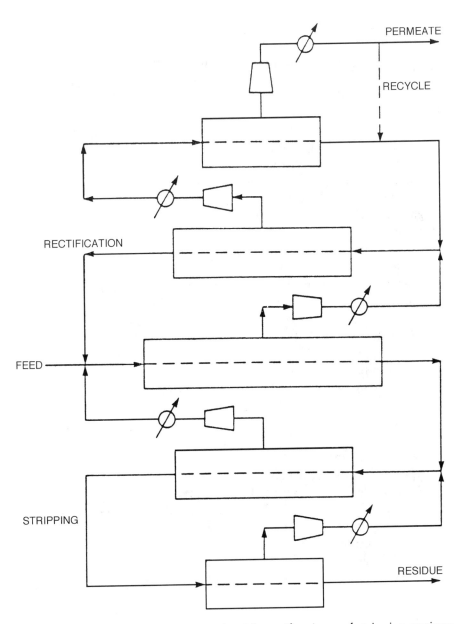

Figure 1.15 Staged permeation cascade with rectification and stripping sections. The individual membrane modules may be operated concurrently or countercurrently or perfect mixing may be assumed to occur. (Source: Bravo et al.[6])

removed to market and utilize the gas. The three principal nonhydrocarbons found are nitrogen, carbon dioxide, and hydrogen sulfide, in that order. Carbon dioxide and hydrogen sulfide are selectively removed by well-known and successful technologies. Chief among these are acid gas absorption and adsorption methods, which are well documented in the literature.

It may be added, however, that membrane systems have been used successfully to separate carbon dioxide from natural gas, notably in enhanced oil recovery operations.[39] Here, (supercritical) carbon dioxide is injected into a petroleum-bearing formation, where the carbon dioxide increases the oil mobility and its subsequent recovery. The carbon dioxide-rich gaseous effluent is recovered, and the carbon dioxide concentrated and reinjected.

With the nitrogen content of subquality natural gas, it is another story. The two principal methods in current but limited use are low-temperature, or cryogenic separation or distillation, and selective adsorption. The former is judged too costly; the latter is starting to make inroads. Membrane separations wait in the wings. More on the general subject of upgrading natural gas follows.

To be adjudged pipeline-quality natural gas, the hydrogen sulfide content must be below 25 grains per SCF (standard cubic foot), which calculates out to about 0.0004 mol %. The hydrogen sulfide removed and recovered may be oxidized to the sulfur oxides, to be vented or, preferably, converted, say, in a lime-water wash for disposal as calcium sulfate (gypsum). In sufficient quantities and concentrations, the recovered hydrogen sulfide may be partially oxidized to elemental sulfur via the Claus process or its equivalent.

Permissible carbon dioxide levels in pipeline-quality natural gas are characteristically up to 2–3 mol %. The recovered carbon dioxide is being increasingly touted for enhanced oil recovery operations rather than being vented to the atmosphere.

The allowable nitrogen content is dictated mostly by the required Btu content for the natural gas. Assuming the natural gas per se at about 1000 Btu/SCF, the nitrogen content could range up to 10 mol %, whereby the Btu content would be no lower than 900 Btu/SCF, the generally accepted cutoff for the Btu-rating. However, pipeline requirements are starting to be more stringent for the nitrogen content and, in some instances, about 3 mol % is the maximum allowable.

Whereas low-temperature or cryogenic methods can be used to separate out the nitrogen, this technology is expensive and not commonly used. The use of selective adsorbents is emerging, and may prove

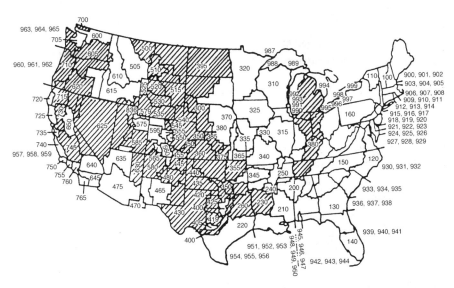

Figure 1.16 Map of subquality natural gas reserves in the United States. (Source: B. J. Moore and S. Stigler.[41]).

economically viable. There is the possibility, however, that membrane separations may prove equally viable, an assessment yet to be determined, and will in large part depend on the further development of suitable membrane materials.

Hence, there are needs, at least on the horizon, to reduce the nitrogen content of natural gases, where in fact perhaps a fourth of the total natural gas reserves can be judged as subquality. Figure 1.16 shows the distribution of subquality reserves in the United States.[9,38] The principal source of the information is the extensive data compilation of the U.S. Bureau of Mines, *Analyses of Natural Gases*, by B. J. Moore and S. Stigler.[40,41] Of more than usual interest is a band of high-nitrogen gas running from southwest Arkansas, across north Texas, into west Texas and the Panhandle, through eastern and northeastern New Mexico, and up into eastern Colorado, then back into western Kansas and down into north-central Oklahoma, virtually completing the circuit. Other notable occurrences are in the Central Valley of California and in West Virginia.

Subquality natural gas is apparently a ready resource, awaiting the need and the necessary upgrading technologies, of which membrane separation is one of the emerging possibilities.

1.5 REPRESENTATIONS AND CALCULATIONS

In many respects, single-stage and multistage membrane separations can be viewed as analogous to the steady-state flash separations and distillations calculated for vapor-liquid systems. The concept of phase equilibrium and the K-value or equilibrium vaporization ratio is replaced by the use of the membrane rate equation, assuming perfect mixing of the respective phases involved. Usually, all these phases are gaseous, but the methodology may be applied to gas-liquid or vapor-liquid systems, or to all-liquid systems.

The single-stage membrane unit becomes equivalent to a so-called flash vaporization. The flash vaporization calculation itself is straightforward, with the vapor and liquid phases assumed at equilibrium, and is presented in a number of references.[42-45] The limits correspond to the dew-point and bubble-point calculations for vapor-liquid equilibrium, which are special or limiting cases for the flash vaporization calculation. It is the object, therefore, to adapt the membrane calculation to the techniques for the flash vaporization calculation and thereby take advantage of the relative simplicity of the latter.

Other procedures and calculation techniques have been developed for both stagewise and differential permeation, such as those presented S-T Hwang and K. Kammermeyer,[46] but they are not pursued here, inasmuch as the analogy is to be made specific to vapor-liquid mass transfer unit operations. In this way, the conventions and techniques already developed for mass transfer operations can be more readily utilized. Also note that the symbols and terminology used for membrane permeation have evolved through the years and vary from one author to another.

Extension can be made to cascade or multistage operations, carried out analogously to a distillation column. The absorption and stripping factor concept or rearrangement may be employed for the calculations, as has also been employed for distillation.[42] The temperature, moreover, is assumed constant from stage to stage. This is the common assumption in absorption, for instance, although in distillation the temperature markedly varies from stage to stage and plate to plate.

The preceding assumes that perfect mixing takes place; that is, there are no changes in flow rate and composition across or perpendicular to the face of the membrane surfaces. Furthermore, the flow rates and compositions at the cell are those of the streams leaving the cell.

Of special consideration is the investigation of the cell as a continuum, first with point withdrawal of the permeate, then in both concurrent and countercurrent flow for the permeate and the reject phases. For this treatment, differential permeation is the mode of attack. The differential forms so obtained must be integrated, however, which becomes a complicating procedure in concurrent flow, and particularly in countercurrent flow. Hence, there is occasion for simplification or reduction.

Note that these types of process calculations, aimed at predicting behavior from known or assigned membrane permeabilities, are not ordinarily pursued in the various updates on the utilization of membrane separations. The topic of utilization can be further searched on the Internet or, more succinctly, in WorldCat, the compendium of books published in the English language, by key word(s) or subject, author, and title. At this writing, a search of WorldCat, for instance, under the title "Membrane Separations..." yields less than 100 titles, whereas a search under the more general title "Membranes" yields results numbering in the thousands. For example, a couple of updates about applications are Donald R. Paul's *Membrane Technology for Gas Separations*,[47] and S. P. Nunes and K.-N. Peinemann's *Membrane Technology in the Chemical Industry*.[48] The field is expanding, as a further search of the literature will reveal. What has been missing are the concomitant calculation procedures, which this volume addresses.

1.6 PERMEATION UNITS

The following derivations and calculations avoid any specific assumptions about units. Instead, the quantities have been described as having "consistent" units. The degree of separation and recovery, moreover, is independent of the units. If, however, numerical values for permeability, pressure, and flow rate are used, then specific units are required, which in turn determine membrane area, given the membrane thickness, or the overall membrane permeability.

Commonly used units for the membrane permeability P_i for a component i are

$$\frac{10^{-9}\,\mathrm{cm}^3}{\mathrm{cm}^2\text{-sec-cm Hg/cm}}$$

that is, 10^{-9} cubic centimeters per square centimeter per centimeter of mercury pressure change per centimeter. The units designated centimeters of mercury (Hg) per centimeter, therefore, pertain to the pressure gradient across the membrane.

Alternately, these units are expressible as

$$\frac{10^{-9}\,\mathrm{cm}^2}{\mathrm{sec\text{-}cm\ Hg}}$$

which, when multiplied by the operating pressure difference in the appropriate and consistent units, becomes the diffusion coefficient or diffusivity; here, this would be measured in 10^{-9} cm^2/sec.

Other units used for the permeability of solids to gases are presented in the *International Critical Tables* (vol. V, pp. 76–77):[49]

$$\frac{10^{-6}\ cm^2}{sec\text{-}atm} \quad or \quad \frac{10^{-9}\ cm^2}{sec\text{-}atm}$$

Thus, the former units are used for say gases through metals and rubber, the latter units for gases through glass. To convert to the pressure units of cm Hg, the numerical values in the preceding units would be multiplied by 76 cm Hg/atm. Some representative values for hydrogen, taken from the *International Critical Tables*,[49] are furnished in Table 1.5. All the permeabilities increase with temperature. A side effect of hydrogen is that it can dissolve into the interstices of metals, affecting the ductility and strength, called *hydrogen embrittlement*. Additionally, hydrogen is reactive, notably to impurities in the metal, that is, components or phases other than the pure metallic state itself. A prime example of this is with steels and their makeup.

Table 1.5 Permeability of Solids to Hydrogen at Elevated Temperatures

System	Temp. (°C)	10^{-6} cm^2/ sec-atm	10^{-9} cm^2/ sec-atm	10^{-9} cm^2/ sec-cm Hg
H-Cu	500	3.5		46.1
	750	8		105
H-Fe	500	100		1316
	600	336		4421
H-Ni	500	3.8		50
	750	31.6		416
H-Pd	500	4450		58550
	600	5750		75660
H-Pt	600	0.77		0.1
	800	4.8		63.2
H-Zn	300	0.4		5.3
H-rubber	20	0.3		3.9
	100	2.6		34.2
H-SiO$_2$	500		6.2–28	0.08–0.37
	800		35–100	0.46–1.3
H-Pyrex	600		Inappreciable	Inappreciable

Source: *International Critical Tables*, vol. V, pp. 76–77.

Henceforth, the symbol P_i or, say, P_j is used for the permeability of an arbitrary component i or j. That is, both i and j denote the key components for the separation, especially for a two-component system but also in a multicomponent system. Inasmuch as the symbol P also is used to denote pressure, as is the common practice, some other symbol could be adopted for permeability, say lower-case, script, or boldface P or Greek or the like. However, the subscripted P_i or P_j seems self-evident, so pressure is also subscripted to provide its distinguishing feature.

Therefore, P_L stands for the higher or upstream pressure (or reject pressure) at the membrane and P_V, for the lower or downstream pressure (or permeate pressure), with the difference $P_L - P_V$ denoting the pressure-drop across the membrane proper. The analogy is akin to that used for phase separations. Furthermore, the usage in the main appears self-evident in context.

The flow is stated in cubic centimeters at standard conditions of temperature and pressure; that is, in standard cubic centimeters. The emphasis here, moreover, is for gases rather than liquids. A statement could be made equivalently, however, for the flow of liquids through membrane barriers.

Overall permeability can be specified by dividing by the membrane thickness Δm in units of

$$\frac{10^{-9}\,\text{cm}^3}{\text{cm}^2\text{-sec-cm Hg}}$$

where cm Hg designates the pressure drop across the membrane in centimeters of mercury. By multiplying by a linear dimension of the membrane, an overall permeability would be measured as

$$\frac{10^{-9}\,\text{cm}^3}{\text{cm-sec-cm Hg}}$$

If, instead, multiplied by the total area of the membrane, an overall permeability would be measured as

$$\frac{10^{-9}\,\text{cm}^2}{\text{sec-cm Hg}}$$

It is understood that this relationship is based on units of cross-sectional area normal to the flow.

The area of the membrane may be designated as the interfacial area and may be measured either as the inner or outer surface of the membrane or as the mean or average, as in the case of a hollow membrane. Thus, a mean value can also be adapted, in common with the practice for the conduction of heat through a tube wall.

The permeability as measured in the preceding centimeter-gram-second (cgs) system may also be converted to the English system, even to the units of darcies or millidarcies for the flow of fluids through porous media, as employed in the petroleum industry.

The concept of the diffusion coefficient, or diffusivity, also is used as the measure of the diffusion of one substance through another. The driving force is the concentration difference, via Fick's law, which can be related to the partial pressure difference by means of the equation of state. It is applicable to mixtures occurring as a common phase but can be applied as well to the case where the second substance is a solid, say, as in dialysis. The dimensions are ordinarily $(distance)^2/time$ (e.g., cm^2/sec). Since the equation of state is implicitly involved, the conversion between permeability and diffusivity is more pronouncedly temperature (and pressure) dependent.

Finally, inasmuch as it is usually more straightforward to work in moles and mole fraction compositions when material balances are involved, it is preferable to put the permeability on a molar basis. The relative permeability, one component to another, is also called the *selectivity* α. In the customary notation used,

$$\alpha = \alpha_{i-j} = P_i/P_j$$

where α_{i-j} is the permeability of component i relative to component j, which is related to the degree of separation that may be attained.

In the derivations and examples that follow, the permeability symbol P_i refers to any of the aforementioned classifications of units, but its particular units are in the context of its usage. For most purposes, moreover, the permeability is an overall or mean permeability.

A few examples are utilized to show various calculations for the conversions between units, as previously indicated.

More data is supplied in Appendix 1.

Membrane Areas for Mixtures

Interestingly, membrane permeability is generally specified for the pure component as such, even though permeabilities for mixtures have been found to be less than for the pure components, as is discussed in Chapter 2.

Consider, therefore, a mixed permeate phase V, which also signifies the molar flow rate. If the mole fraction of a component i present in the permeate phase is designated y_i and the mole fraction present in the feed-reject phase L is denoted x_i, then the permeation relationship should presumably be of the basic form

$$Vy_i = P_i(P_L x_i - P_V y_i)A$$

where A is the membrane interfacial area and $(P_L - P_V)$ is the total pressure difference across the membrane proper. However, the partial pressure difference $(P_L x_i - P_V y_i)$ represents the driving force for component i, so to speak, albeit it may be adjusted using the idea of an "activity" difference.

Moreover, the units should be consistent; that is, the permeability P_i should have the units of moles per unit time per unit area per unit pressure (or partial pressure) difference and refer to the permeability measured for the pure component i. As such, it would denote an overall permeability; that is, the membrane thickness has already been taken into account.

The preceding is the basic form adapted for Chapter 2 and the following chapters, albeit the membrane area A may be incorporated into the permeability term for simplification purposes.

Assuming, however, that the permeability is represented in the units of moles/area-time-pressure difference, then the membrane area would be calculated from

$$A = \frac{Vy_i}{P_i(P_L x_i - P_V y_i)}$$

Again, the units should be consistent. For the permeation of a pure component only, this would yield the expected relationship, where y_i and x_i are unity.

Finally, it is emphasized that the subsequent membrane separation derivations and calculations involving two or more components should be—and are—internally consistent. That is to say, for the purposes here, the same membrane area requirement results whether we deal with component i, component j, or whichever component of the feed mixture. In other words, the equation derivations and calculations are to be perceived simultaneously for each component, one to another.

EXAMPLE 1.1

A membrane has a nominal permeability of 20 in the standard or customary units of $(10^{-9})cm^3/cm^2$-cm Hg/cm, and a thickness of 10 μ, or 10 microns, or 10×10^{-6} meters, or 10×10^{-3} millimeters, or $10 \times$

$10^{-4} = 10^{-5}$ centimeters. Perhaps a more useful conversion is as follows:

$$P_i = \frac{20(10^{-9})}{10(10^{-4})}\frac{76}{22,414} = 20(0.00339)(10^{-6})\frac{\text{g-moles of } i}{\text{cm}^2\text{-sec-atm}}$$

Alternately,

$$P_i = 20(0.00339)(10^{-6})(30.48)^2\,\frac{1}{453.59}\,\frac{1}{14.696}$$

$$= 20(0.000472)(10^{-6})\frac{\text{lb-moles of } i}{\text{ft}^2\text{-sec-psi}}$$

$$= 20(1.700)(10^{-6})\frac{\text{lb-moles of } i}{\text{ft}^2\text{-hr-psi}}$$

The particular values and units used depend on the circumstance.

EXAMPLE 1.2

Apropos of Example 1.1, a membrane cell is to have the following characteristics:

$$P_i = 20(10^{-9})\ \text{cm}^3/\text{cm}^2\text{-sec-cm Hg/cm}$$

$$P_j = 10(10^{-9})\ \text{cm}^3/\text{cm}^2\text{-sec-cm Hg/cm}$$

The pressure P_L on the high-pressure or reject side of the membrane and the pressure P_V on the low-pressure or permeate side of the membrane are to be as follows:

$$P_L = 3(10^1)\ \text{atm} \qquad \text{or} \qquad 30\ \text{atm}$$

$$P_V = 2(10^1)\ \text{atm} \qquad \text{or} \qquad 20\ \text{atm}$$

For a membrane thickness of 10 μ (10 microns), the permeabilities convert to

$$P_i = 20(76/22,414)(10^{-6})\ \text{g-moles/cm}^2\text{-sec-atm}$$

$$P_j = 10(76/22,414)(10^{-6})\ \text{g-moles/cm}^2\text{-sec-atm}$$

where $76/22,414 = 0.00339$. The product of the permeabilities in these dimensions times the pressure or pressure difference in atm yields an equivalent flux rate in g-moles/cm²-sec.

Anticipating the problem statement and results as per Example 3.1, the overall permeability and the pressures may be assigned arbitrary or unspecified dimensions such that

$$P_i = 20 \qquad P_L = 3$$

$$P_j = 10 \qquad P_V = 2$$

For the arbitrary characteristics listed, an arbitrary permeate rate is determined in Example 3.1 with the value $V = 12.7056$ for an assigned permeate-to-feed (V/F) ratio of 0.5, and for component i has the dimensions of overall permeability times pressure ($P_i P_V$); that is, it is a permeate flux rate. In other words, considering the expression for the dimensionless K-value, developed in Chapter 3, if

$$K_i = \frac{P_i P_L}{V + P_i P_V}$$

then the appropriate dimensions can be introduced as a multiplier into both the numerator and denominator:

$$K_i = \frac{[20(3)](10^{-9})\dfrac{1}{10(10^{-4})}\dfrac{76}{22{,}414}(10)}{V(10^{-9})\dfrac{1}{10(10^{-4})}\dfrac{76}{22{,}414}(10) + [20(2)](10^{-9})\dfrac{1}{10(10^{-4})}\dfrac{76}{22{,}414}(10)}$$

where $V = 12.7056$ and $[20(3)]$ and $[20(2)]$ are the original values of $P_i P_L$ and $P_i P_V$, originally with arbitrary dimensions, as in Example 3.1. (K_i calculates to 1.138399.)

To continue, on introducing the dimensions, the value of the total permeate flux rate becomes

$$(12.7056)(76/22{,}414)(10^{-9}/10^{-3})(10) = 0.4308(10^{-6}) \text{ g-moles/cm}^2\text{-sec}$$

If the feed rate is, say, 1 g-mole/sec and the permeate to feed ratio is 0.5, then the membrane area requirement, in turn, becomes

$$\text{Area} = \frac{1(0.5)}{0.4308(10^{-6})} = 1.16(10^6) \text{ cm}^2$$

Similar conversions can be made to other units.

EXAMPLE 1.3

The unit of permeability used in the petroleum industry for the flow of fluids through porous media, in oil and gas production, is the darcy.[44,45] Usually, the permeability of oil- and gas-producing formations is given in millidarcies, a millidarcy being 1/1000 of a darcy.

The origins are in the Darcy (d'Arcy) relationship, which relates flow rate to the pressure gradient:

$$v = -\frac{K}{\mu}\frac{dP}{dx}$$

where

v = superficial flow velocity;
K = permeability coefficient for flow through porous media;
μ = viscosity;
P = pressure;
x = linear dimension opposite to the direction of flow.

The superficial velocity is the actual volumetric rate divided by the total cross-sectional area normal to the direction of flow.

A porous medium having a permeability of 1 darcy, at standard conditions, permits the flow of a fluid of 1 centipoise viscosity at the superficial rate of 1 cm/sec under a pressure gradient of 1 atm/cm. In this formula, v is in cm/sec, K is in darcies, μ is in centipoises (centigrams/cm-sec), P is in atm, and x is in cm. (A centipoise is 1/100 of a poise. A poise has the dimensions of g/cm-sec, whereas a centipoise has the dimensions of centigrams/cm-sec.) The actual units for K in darcies would be centigram-cm/atm-sec^2.

In the English system, the superficial flow velocity is in ft/sec, the permeability coefficient is in ft^3/sec^2, the viscosity is in British viscosity units (BVUs) of lb/ft-sec (to convert to BVU, multiply the viscosity in centipoises by 6.72×10^{-4}), P is in psf (lb per ft^2), and x is in ft.

If the flow velocity is to be in ft/hr, then the viscosity is measured in lb/ft-hr; that is, the viscosity in centipoises is multiplied by $(6.72 \times 10^{-4})(3600) = 2.42$.

Conversion may be made to the actual volumetric rate by multiplying by the cross-sectional area normal to flow. In turn, conversion may be made to the mass flow rate by multiplying the actual volumetric flow rate by the density of the fluid at flow conditions. Dividing by the molecular weight of the fluid will give the molar flow rate.

The conversion between K in say ft^3/hr^2 and K' in darcies (or darcys) is given by

$$K\,(\text{in ft}^3/\text{hr}^2) = K'\,(\text{in cg-cm/atm-sec}) \times \dfrac{\dfrac{1}{100}\dfrac{1}{453.6}\dfrac{1}{30.48}}{(14.696)(144)\left(\dfrac{1}{3600}\right)}$$

$$= K'\,(\text{in cg-cm/atm-sec}) \times 0.00443$$

that is,

$$K\,(\text{in ft}^3/\text{hr}^2) = K'\,(\text{in darcies}) \times 0.00443$$

To obtain K in ft^3/sec^2, the conversion is

$$K\,(\text{in ft}^3/\text{sec}^2) = K'\,(\text{in darcies}) \times (1/3600)^2 \times 0.00443$$

$$= K'\,(\text{in darcies}) \times 3.418(10^{-10})$$

The ratio K/μ is called the *mobility*. Based on the Darcy concept, the mobility has the dimensions of cm^2/sec-atm. The mobility for porous media is, in fact, identical to what is called the *permeability* as used in the nomenclature for membrane permeation. The conversion of units is, therefore, of interest:

$$\frac{K}{\mu}\left(\text{in } \frac{\text{cm}^2}{\text{sec-atm}}\right) = P_i\left(\text{in } \frac{10^{-9}\ \text{cm}^3}{\text{cm}^2\text{-sec-cm Hg/cm}}\right) \times 76 \times 10^{-9}$$

$$\frac{K}{\mu}\left(\text{in } \frac{\text{darcies}}{\text{centipoises}}\right) = P_i\left(\text{in } \frac{10^{-9}\ \text{cm}^3}{\text{cm}^2\text{-sec-cm Hg/cm}}\right) \times 76 \times 10^{-9}$$

This value of K is properly subscripted K_i to denote the permeability to a component i or I. As an example, if the membrane permeability were

$$P_i = 20\ (10^{-9})\ \text{cm}^3/\text{cm}^2\text{-sec-cm Hg/cm}$$

then the mobility would be

$$K/\mu = 20(10^{-9})(76) = 1.520(10^{-6})\ \text{darcies/centipoises or cm}^2/\text{sec-atm}$$

For a gas with a viscosity of 0.01 centipoise, the corresponding permeability coefficient K for component i is

$$K_i = 1.520(10^{-8})\ \text{darcies or } 1.520(10^{-5})\ \text{millidarcies}$$

This value is extremely low, since a typical permeable petroleum reservoir formation or rock may have a permeability on the order of as much as 1 darcy or 1000 millidarcies or as little as, say, 10^{-2} darcies or 10 millidarcies, or conceivably even 10^{-3} darcies or 1 millidarcy, but that is still many orders of magnitude greater than membrane permeability or mobility.

REFERENCES

1. Lonsdale, H. K. "The Growth of Membrane Technology." *Journal of Membrane Science* 10 (1982), pp. 81–181.
2. Flynn, Thomas M., and J. Douglas Way. *Membrane Separations in Chemical Processing*. Boulder CO: U.S. Department of Commerce, National Bureau of Standards, 1982.
3. Gas Research Institute. *Proceedings of the First GRI Gas Separations Workshop*, held in Denver, October 22–23, 1981. Chicago: Gas Research Institute, 1982.
4. Gas Research Institute. *Proceedings of the Second GRI Gas Separations Workshop*, held in Boulder, CO, October 21–22, 1982. Chicago: Gas Research Institute, 1983.
5. Signal Research Center for the U.S. Department of Energy. *Membrane Separation Processes in the Petrochemical Industry*, Phase II, April 1, 1984–October 30, 1985. Report No. DOE/ID/12442-1. Washington, DC: U.S. Department of Energy, 1985.
6. Bravo, Joseph L., James R. Fair, Jimmy L. Humphrey, Chris L. Martin, Albert F. Seibert, and Sudhir Joshi. *Membrane Separations Technology Review, in Assessment of Potential Energy Savings in Fluid Separation Technologies: Technology Review and Recommended Research Areas. Final Report*. Report No. DOE/ID/122473-1 (DE85013839). Contract No. AS07-83ID12473. Washington, DC: Technical Information Center, Office of Scientific and Technical Information, U.S. Department of Energy, December 31, 1984.
7. Leeper, Stephen A., Daniel H. Stevenson, Peter Y.-C. Chiu, Stephen J. Priebe, Herbert F. Sanchez, and Penny M. Wikoff. *Membrane Technology and Applications: An Assessment*. Report No. EGG-2282 (DE84009000), prepared for U.S. Department of Energy, Idaho Operations Office under DOE Contract No. DE-AC07-76ID01570. Idaho Falls, ID: EG&G Idaho Inc., February 1984.
8. Committee on Separation Science and Technology, Board on Chemical Science and Technology, Commission on Physical Sciences, Mathematics, and Resources, National Research Council. "Generic Research Frontiers." In *Separation and Purification: Critical Needs and Opportunities*. Washington, DC: National Academy Press, 1987.
9. Hoffman, E. J. "Subquality Natural Gas: The Resource and Its Potential." In *Investigation of Structure Permeability Relationships of Silicone Membranes, Final Report*, ed. Chi-long Lee, S. A. Stern, J. E. Mark, and E. Hoffman. Report No. GRI-87/0037. Chicago: Gas Research Institute, 1987. Report available from the National Technical Information Service.

10. Hoffman, E. J., K. Venkataraman, and J. L. Cox. "Membrane Separations for Subquality Natural Gas." *Energy Progress* 8, no. 1 (March 1988), pp. 6–13.
11. Kamalesh K. Sirkar and Douglas R. Lloyd, eds. *New Membrane Materials and Processes for Separation.* AIChE Symposium Series, no. 261, vol. 84. 1987 AIChE Summer National Meeting, Minneapolis. New York: American Institute of Chemical Engineers, 1988.
12. Govind, R., and N. Itoh, eds. *Membrane Reactor Technology.* AIChE Symposium Series, no. 268, vol. 85. AIChE Annual Meeting, Washington, DC, November 1988. New York: American Institute of Chemical Engineers, 1989.
13. Fouda, A. E., J. D. Hazlett, T. Matsuura, and J. Johnson, eds.; C. R. Bartels, B. S. Minhas, and B. Ryan, co-eds. *Membrane Separations in Chemical Engineering.* AIChE Symposium Series, no. 272, vol. 85. 1989 AIChE Spring National Meeting, Houston, April 2–6, 1989. New York: American Institute of Chemical Engineers, 1989.
14. João G. Crespo and Karl W. Böddeker, eds. *Membrane Processes in Separation and Purification.* Dordrecht, the Netherlands: Kluwer Academic Publishers, 1993. Published in cooperation with NATO Scientific Affairs Division.
15. Paul, D. R., and Yuri P. Yampol'skii, eds. *Polymeric Gas Separation Membranes.* Boca Raton, FL: CRC Press, 1994.
16. Prasad, R., R. L. Shaner, and K. J. Doshi. "Comparison of Membranes with Other Gas Separation Technologies." In *Polymeric Gas Separation Membranes,* ed. D. R. Paul and Yuri P. Yampol'skii. Boca Raton, FL: CRC Press, 1994.
17. Richard D. Noble and J. Douglas Way, eds. *Membrane Separations Technology: Principles and Applications.* Amsterdam, the Netherlands: Elsevier Science, 1995.
18. Richard D. Noble and J. Douglas Way, eds. *Liquid Membranes: Theory and Applications,* developed from a symposium presented at the Eighth Rocky Mountain Regional Meeting of the American Chemical Society in Denver, CO, June 8–12, 1986. Washington, DC: American Chemical Society, 1987.
19. Anna Crull, ed. *Membrane and Separations Technology Industry Review.* Norwalk, CT: Business Communications Company, 1998.
20. Baker, R. W., E. L. Cussler, W. Eykamp, W. J. Koros, R. L. Riley, and H. Strathmann. *Membrane Separation Systems: Recent Developments and Future Directions.* Park Ridge, NJ: Noyes Data Corp., 1991.
21. Garcia, Fred. *Review of Membrane Technology for Methane Recovery from Mining Operations.* Pittsburgh: U.S. Department of the Interior, Bureau of Mines, 1988.
22. *Silica Membranes for Hydrogen Separation from Coal Gas,* Quarterly Progress Report, December 1, 1992–February 28, 1993, California Institute of Technology, Pasadena, 1993. Report No. DOEPC92525T2. Contract No. FG2292PC92525. Washington, DC: U.S. Department of Energy, April 15, 1993.
23. *Gas Separations Using Ceramic Membranes, Final Report, September 1988–February 1993,* prepared for U.S. Department of Energy. Report No. DOEMC251353341, Contract No. AC2188MC25135. Alcoa Center, PA: Aluminum Co. of America, February 1993.
24. *High Temperature Membranes for H_2S and SO_2 Separations,* Quarterly Progress Report, October 1, 1992–December 31, 1992. U.S. Department of Energy. Report No. DOEPC9029318, Contract No. FG2290PC90293. Atlanta: School of Chemical Engineering, Georgia Institute of Technology, 1993.

25. *Upgrading of Raw Natural Gas Using High-Performance Polymer Membranes.* Report No. DOEMETCC937068, CONF930951662. Morgantown, WV: Department of Energy, Morgantown Energy Technology Center, 1993.

26. Department of Chemical Engineering, University of Southern California, Los Angeles. *High Temperature Ceramic Membrane Reactors for Coal Liquid Upgrading, Final Report, September 21, 1989–November 20, 1992.* Report No. DOEPC89879T13, Contract No. AAC2289PC89679. Washington, DC: U.S. Department of Energy, 1992.

27. Department of Chemical Engineering, University of Colorado, Boulder. *Direct Methane Conversion to Methanol, Quarterly Status Report, October 1, 1992–December 31, 1992.* Report No. DOEMC271153343, Contract No. FG2190MC27115. Washington, DC: U.S. Department of Energy, 1993.

28. Cabasso, I., S. A. Stern, H. R. Acharya, W. Li, E. Korngold, T. Makensie, Z. Liu, and E. Poda. *Energy Efficient Membrane Processes for the Separation of Organic Liquids.* Report No. DOE/ID/12232-1-P1,P2,P3. Washington, DC: U.S. Department of Energy, 1987.

29. Leeper, S. A., and G. T. Tsao. *Membrane Separation in Ethanol Recovery: An Analysis of Two Applications of Hyperfiltration.* Report No. EGG-J-05285. Idaho Falls, ID: EG&G Idaho Inc., 1986.

30. W. S. Winston Ho and K. K. Sirkar, eds. *Membrane Handbook.* New York: Van Nostrand, 1992.

31. Noble, R. D., and S. A. Stern, eds. Membrane Science and Technology Series. *Membrane Separations Technology: Principles and Applications.* Amsterdam, the Netherlands: Elsevier, 1995.

32. McKetta, J. J., and W. A. Cunningham, eds. *Encyclopedia of Chemical Processing and Design,* vol. 27. New York: Marcel Dekker, 1988.

33. Grayson, M., and D. Eckroth, eds. *Kirk-Othmer Encyclopedia of Chemical Technology,* 3d ed., vol. 15. New York: Wiley, 1981.

34. Perry, Robert H., late ed.; Don W. Green, ed.; James O. Maloney, assoc. ed. *Perry's Chemical Engineers' Handbook,* 7th ed. New York: McGraw-Hill, 1997.

35. Mazur, W. H., and M. C. Chan. "Membranes for Natural Gas Sweetening and CO_2 Enrichment." *Chemical Engineering Progress* 78, no. 10 (1982), p. 38.

36. Schell, W. J. "Membrane Use Technology Growing." *Hydrocarbon Processing* 62, no. 8 (1983), p. 43.

37. Schell, W. J., and C. D. Houston. "Spiral-Wound Permeators for Purification and Recovery." *Chemical Engineering Progress* 78, no. 10 (1982), p. 33.

38. Hoffman, E. J. "Subquality Natural Gas Reserves." *Energy Sources* 10, no. 4 (1988), pp. 239–245.

39. Coady, A. B., and J. A. Davis. "CO_2 Recovery by Gas Permeation.," *Chemical Engineering Progress* 78, no. 10 (1982), p. 44.

40. Moore, B. J. *Analyses of Natural Gases, 1917–1980,* Bureau of Mines Information Circular 8870. Washington, DC: U.S. Department of the Interior, 1982.

41. Moore, B. J., and S. Stigler. *Analyses of Natural Gases, 1917–85,* Bureau of Mines Information Circular 9129. Washington, DC: U.S. Department of the Interior, 1986, updated yearly; also available from NTIS on magnetic tape, floppy disks, or high-density disks.

42. Hoffman, E. J. *Azeotropic and Extractive Distillation*. New York: Wiley Interscience,1964; Huntington, NY: Krieger, 1977.
43. Brown, G. G., A. S. Foust, D. L. Katz, R. Schneidewind, R. R. White, W. P. Wood, G. M. Brown, L. E. Brownell, J. J. Martin, G. B. Williams, J. T. Banchero, and J. L. York. *Unit Operations*. New York: Wiley, 1950.
44. Katz, D. L., D. Cornell, R. Kobayashi, F. H. Poettmann, J. A. Vary, J. R. Elenbaas, and C. F. Weinaug. *Handbook of Natural Gas Engineering*. New York: McGraw-Hill, 1959.
45. Hoffman, E. J. *Phase and Flow Behavior in Petroleum Production*. Laramie, WY: Energon, 1981.
46. Hwang, S.-T., and K. Kammermeyer. *Membranes in Separations*, vol. VII of *Techniques of Chemistry*, ed. A. Weissberger, Chapters 4 and 8, Appendix B. New York: Wiley Interscience, 1975.
47. Paul, Donald R. *Membrane Technology for Gas Separations*. New York: American Institute of Chemical Engineers, 1999.
48. Nunes, S. P., and K.-N. Peinemann, eds. *Membrane Technology in the Chemical Industry*. New York: Wiley, 2001.
49. *International Critical Tables*. New York: McGraw-Hill, 1926–1930.

2

Membrane Permeation Relationships

In many respects, single-stage and multistage membrane separations can be viewed as analogous to steady-state flash separations and distillations, as derived and calculated for vapor-liquid systems. By a rearrangement of the membrane permeation rate or flux equation, the result is adaptable to the form used for vapor-liquid phase equilibrium, expressed as the K-value or equilibrium vaporization ratio. Therefore, in the usual notation, assuming perfect mixing in the respective membrane phases involved, $y_i = K_i x_i$, where K_i is the ratio of the mole fraction of a component in the permeate phase with respect to the mole fraction in the reject or retentate phase. This derivational sequence, detailed in Chapter 3, is basically connected with the idea of permeability in its several embodiments or units and the corresponding rate or flux relationships, which constitute the subject of this chapter.

Both the permeate and reject phases, for the most part, are considered gaseous for the baseline derivations and calculations, but the methodology is equally applicable to gas-liquid or vapor-liquid systems, as in pervaporation, and to miscible liquid-liquid systems and solutions of dissolved solids, even to colloids, suspensions, and emulsions. All that is required is a mathematical conversion of permeability units and values, along with component concentrations, to be consistent with the gas-phase format, which is expressed in terms of mole fractions and their partial-pressure difference as the driving force for permeation.

To continue, for liquid-liquid phase equilibria per se, we may speak more generally of a distribution coefficient (also denotable as K or K_i) for relating the component mole fraction composition in one liquid phase to that in the other. Albeit concentrations constitute the usual mode of expression for liquids, concentrations are transformable to mole fractions via the equation of state as applied to a liquid or a more-dense single phase. However, the membrane permeation rate or flux balance modifies into the same format as for gaseous systems to yield the K-value form for relating the permeate composition to the reject composition. All that is required is the conversion of permeability (or diffusivity) into the proper and consistent units and of concentrations into mole fractions. (For a pure component or mixture of constant composition, liquid compressibility can be used to convert concentrations—that is, densities—to pressure.)

The single-stage membrane unit becomes equivalent to a so-called flash vaporization. The flash vaporization calculation itself is straightforward, with the vapor and liquid phases assumed at equilibrium, as presented in a number of references.[1-4] The limits correspond to the dew-point and bubble-point calculations for vapor-liquid equilibrium, which are special or limiting cases for the flash vaporization calculation. The object, therefore, is to adapt the membrane calculation to the techniques for the flash vaporization calculation and thereby take advantage of the relative simplicity of the latter, as is demonstrated in Chapter 3.

Other procedures and calculation techniques have been developed for both stagewise and differential permeation, as are presented by S.-T. Hwang and K. Kammermeyer,[5] but these are not pursued here, inasmuch as the analogy is to be made specific to vapor-liquid mass transfer unit operations. In this way, the conventions and techniques already developed for mass transfer operations can be more readily utilized. Note also that the symbols and terminology used for membrane permeation have evolved through the years and perforce vary from one author to another.

In "Process Design and Optimization," R. Rautenbach explores the uses of cascade operations, without and with reflux, and of operations as a continuum.[6] These and other embodiments, and their modes of operation and calculation, are detailed in subsequent chapters.

In further preview, extension can be made to cascade or multistage operations, provided in Chapter 4, and carried out analogously to a distillation column. The absorption and stripping factor concept or rearrangement can be employed for the derivations and calculations, as done for distillation.[1] The temperature, moreover, is appropriately assumed constant from stage to stage. This is the common assumption in absorption, for instance, although in distillation the temperature varies from stage to stage and plate to plate.

The preceding assumes that perfect mixing takes place; that is, no changes in flow rate and composition across or perpendicular to the face of the membrane surfaces. Furthermore, the flow rates and compositions at the cell are to be that of the streams leaving the cell.

Of special consideration is first to examine the cell as a continuum, then with the point withdrawal of permeate, and finally for both concurrent and countercurrent flow for the permeate and the reject phases. These are the subjects respectively for Chapters 5, 6, and 7. For this alternative treatment, differential permeation is the mode of attack. The differential forms so obtained must be integrated, however, which becomes a complicating procedure in concurrent flow, and particularly for countercurrent flow. Hence, there is occasion for simplification and reduction, which therefore are introduced.

Lastly, Chapter 8 examines the matter of membrane reactors, which pertains to chemically reacting systems, either homogeneous or heterogeneous (catalyzed). The selective removal of one or another of the reaction products by membrane permeation shifts the reaction conversion equilibrium "to the right."

In short, all these embodiments, derivations, and calculations depend on the units and values for the membrane permeation or permeability coefficients and their consistency with the membrane permeation rate or flux relationship. These latter matters, of permeability coefficients and flux relationships, are further addressed in the following sections.

2.1   PERMEATION RATES

The rate of mass transfer for a component i may be said to be proportional to the gradient or difference in some generalized quantity, here called the *absolute activity*, denoted by the symbol γ_i. (Some other symbol can, of course, be used, such as λ_i, but this is preempted by its use for latent heat, the usual problem encountered with more needs than readily available symbols.) Therefore, if the difference or gradient is zero, no net transfer occurs.

Features of Absolute Activity

As just indicated, absolute activity is to be such a function of composition that the rate or flux for the transfer of a component i is directly proportional to ($\propto$) its gradient, or difference. That is, for the unidirectional, steady-state, mass, or molar flux G_i of a component i,

$$G_i \propto -\frac{d\gamma_i}{dx} \quad \text{or} \quad G_i \propto -\frac{\Delta\gamma_i}{\Delta x} \quad \text{or} \quad G_i \propto -\Delta\gamma_i$$

where the x direction is opposite to the transfer direction and the incremental distance may be absorbed into the proportionality constant (or coefficient). Alternately, as per the last proportionality, we may speak of mass transfer across an interface as between phases. (We also have the option of speaking in terms of a volumetric rate or flux, as with a liquid component, or with a gas at, say, standard conditions.)

A further feature of absolute activity is to be that its value in each of two phases is the same if the phases are in equilibrium. Thus, the net rate of mass transfer between phases is zero if the absolute activity difference is zero.

In heat transfer, the temperature serves as such a function, there being no heat transfer within or between phases at a constant temperature. Similarly, in fluid flow, pressure serves such a function, there being no flow at a constant pressure.

For component mass transfer within or between phases, usage requires an explicit formulation, which as a boundary condition may be made compatible with the laws for the flow of a single component or a mixture of constant composition. These relationships form Darcy's (d'Arcy's) law for the flow of fluids through porous media and even Poiseuille's law for viscous flow through cylinders, tubing, or piping. The flow rate, in effect, is proportional to the pressure drop or gradient, depending on whether the fluid is a gas or liquid, which affects the density. This, in turn, affects the representation of the mass or molar flow rate or mass or molar flux, vis-à-vis the volumetric flow rate or volumetric flux (the linear velocity). The conversion between representations can be made via an equation of state—not only for a less-dense phase, called a *gas*, but for a more-dense phase, called a *liquid*. The more usual deployment for the equation of state is in the form of the gas law, whereby a compressibility factor z is introduced to correlate departure from the behavior of a perfect gas and is applicable to what are called *liquids* as well as to gases or vapors.

Diffusion

Of prime interest here is compatibility with the steady-state, unidirectional form for Fick's law of diffusion, which may be expressed as

$$G_i = -D_i \frac{dc_i}{dx}$$

where

G_i = mass or molar flux of a component i;
D_i = the diffusion coefficient or diffusivity for component i;
c_i = concentration of component i in mass or moles per unit volume;
x = linear coordinate in direction opposite to flow.

All the units are to be consistent, and the units for D_i are in (distance)^2/time, commonly cm^2/sec in the centimeter-gram-second (cgs) sytem. The relationship applies to both gases and liquids and pertains to the motion of a component i with respect to its medium. Values for gases are typically around unity in cgs units and, for liquids, around 10^{-5} cm^2/sec or less.

Extension to Permeability and Partial Pressure

The medium may also constitute a membrane, whether solid or liquid; and the motion of a gaseous, liquid, solvated, or ionic component relative to and through this medium is entirely analogous to diffusion, by which

$$G_i = -P_i \frac{dc_i}{dx}$$

where P_i is the permeability coefficient, in the appropriate and consistent units, here in (distance)^2/time, the same as for D_i. The preceding is readily transformed to partial pressure units by introducing the equation of state for the mixture, designable as

$$PV = znRT$$

whereby

$$c_i = \frac{n_i}{V} = \frac{P}{zRT}\frac{n_i}{n} = \frac{Px_i}{zRT} = \frac{pp_i}{zRT}$$

where $n = \Sigma n_i$ is the total number of moles in the mixture, P and T are system pressure and absolute temperature, and R is the gas constant in consistent units. The behavior of the compressibility factor z may vary widely, depending on conditions and whether the mixture is in a gaseous or liquid phase. The symbols $Px_i = pp_i$ are used to denote partial pressure, where here x_i is the mole fraction in the phase under consideration, whether gas or liquid.

Substituting, at constant temperature,

$$G_i = -\frac{P_i}{RT}\frac{d\left(\dfrac{Px_i}{z}\right)}{dx}$$

Therefore, for a pure component or mixture of constant composition and depending on the behavior of z, the mass or molar flux within the membrane is proportional to the pressure gradient. For discrete changes across the membrane,

$$G_i = -\frac{P_i}{RT}\frac{\Delta\left(\dfrac{Px_i}{z}\right)}{\Delta m} = -\frac{P_i/\Delta m}{RT}\Delta\left(\frac{Px_i}{z}\right) = -\frac{P_i/\Delta m}{\bar{z}RT}\Delta(Px_i)$$

where Δm is the membrane thickness. The use of a mean or average value of z may or may not be warranted; for example, in pervaporation, where the mixture changes from a liquid to a gas during permeation. The difference $\Delta(Px_i)$ may be adapted to the particular nomenclature and symbolism used, as is subsequently shown.

Phase Equilibrium and Mass Transfer

A problem with concentration (or partial pressure) as used in Fick's law is that the respective component concentrations (or partial pressures) are different within two phases at equilibrium, save in the case of homogeneous azeotropes. Therefore, this phase-equilbrium feature of absolute activity is not apparent. It can be accommodated, however, by assigning an equilibrium absolute activity to the one phase that, by definition, would be equal to that of the other. This accommodation is most apparent in its assignment to the behavior of vapor-liquid systems.

For gases or vapors, the partial pressure is the common substitution for what we call *absolute activity*. Moreover, this is compatible with the idealization for mixtures that combines Raoult's law for partial pressures with Dalton's law; that is,

$$pp_i = Py_i = (vp)_i x_i$$

where

pp_i = partial pressure of component i;
P = total system pressure;
y_i = mole fraction of component i in the vapor phase;
$(vp)_i$ = vapor pressure of pure component i;
x_i = mole fraction of component i in the liquid phase.

In the form expressible as $pp_i = H_i x_i$, the preceding constitutes Henry's law (for dilute mixtures or solutions), where H_i is Henry's constant for component i, as ordinarily obtained by experiment. Ideally, therefore, Henry's constant can be construed as the vapor pressure of the pure component.

By dividing through by total pressure, the K-value form is obtained:

$$y_i = \frac{(vp)_i}{P} x_i = K_i x_i$$

This serves as a sometimes fair approximation for vapor-liquid K-value behavior for mixtures but is modified by experimental phase-equilibrium determinations, notably as systematically obtained for the lower hydrocarbons, and presented as charts or correlations. (This idealization based on Raoult's law can be adjusted to agree more closely with experiment by the insertion of what is called an *activity coefficient*, customarily also designated by the symbol γ_i:

$$y_i = \gamma_i \frac{(vp)_i}{P} x_i = K_i x_i$$

Other terms such as fugacity sometimes enter into the formulations. We choose not to further digress.)

The driving force or absolute activity difference for mass transfer becomes the difference

$$\gamma_i - \gamma_i^* = Py_i - Py_i^* = Py_i - vp_i x_i = Py_i - PK_i x_i$$

where

y_i = mole fraction composition in the gas phase;
y_i^* = mole fraction composition of a vapor-phase "film" that would be in equilibrium with the liquid phase.

The flux equation for the mass (or molar) transfer of component i between phases can therefore be expressed as

$$G_i = k_g P(y_i - y_i^*) = k_g P(y_i - K_i x_i)$$

where k_g may be called the gas-film mass (or molar) transfer coefficient for component i.

For liquid-liquid systems, no such ready simplification is apparent, and experimental data is the rule, albeit the linear approximation sometimes can be used for a distribution coefficient, such as $(K_{A-B})_i$, whereby at equilibrium between two phases A and B,

$$(x_A)_i^* = (K_{A-B})_i (x_B)_i$$

such that

$$G_i = k_L P[(x_A)_i - (x_A)_i^*] = k_L P[(x_A)_i - (K_{A-B})(x_B)_i]$$

where k_L is the liquid-film mass (or molar) transfer coefficient for component i in partial-pressure units. Alternately, concentrations can be substituted by utilizing the equation of state.

For dissolved solids in (dilute) solutions, the distribution of solute between phases can be expressed in similar fashion, at least over limited intervals.

Otherwise, the membrane analogy for Fick's law of diffusion comes into play, whereby the species flow rate or flux is perceived as proportional to a concentration gradient or, overall, to a concentration difference but which is convertible into a partial-pressure gradient or difference.

Consistency of Units

Whatever the relationship used, the particular units for each symbol or entity must be consistent with the units for all other entities, which may or may not involve a conversion of units. Therefore, the units of, say, permeability are defined within the context of the rate or flux relationship used and the units of the driving force; for example, partial pressure or pressure times mole fraction.

Gaseous Systems

Accordingly, from the preceding, the absolute activity of a component i in a vapor or gaseous phase can be conveniently expressed merely in terms of the partial pressure, which is the custom. That is, in terms preferably of mole fractions,

$$\gamma_i = Py_i$$

and the units are in partial pressure. Therefore, we can speak of the mass or molar flux G_i of a component i in the vapor or gas phase as being related by the following rate equation for permeation:

$$G_i = Gy_i = -P_i \frac{d\gamma_i}{dx} = -P_i \frac{d(Py_i)}{dx}$$

where, in this context, P_i is a pointwise coefficient of permeability and the symbol x signifies increasing distance opposite to linear flow.

If $G_i = Gy_i$ is to be the molar flux of a component i, it follows that the units for the pointwise permeability coefficient P_i necessarily are

$$\frac{\text{moles of } i}{\text{time-area-} \dfrac{pp_i}{\text{distance}_i}}$$

where, for convenience, the symbol pp_i denotes Py_i, by definition the partial pressure of component i. Moles, of course, are readily converted to gas volume, for example, at standard conditions. Hence, gaseous permeability can be expressed in terms of volume, say, in standard cc. Note that the mole fraction is generally considered dimensionless, albeit it is actually the moles of, say, a component *i* divided by the total moles.

It may be observed that Fick's law for diffusion as applied to gases is fully equivalent to the preceding. By Fick's law, where the driving force is a concentration difference,

$$G_i = -D_i \frac{dc_i}{dx}$$

where D_i is the diffusivity or diffusion coefficient in consistent units; that is, D_i has the units

$$\frac{\text{moles of } i}{\text{time-area-} \dfrac{\text{moles of } i}{\text{volume-distance}}} = \frac{(\text{distance})^2}{\text{time}}$$

By the perfect gas law, commonly expressed as $PV = nRT$, it follows that

$$c_i = \frac{n_i}{V} = \frac{n_i P}{nRT} = \frac{1}{RT} P y_i$$

where the total number of moles is $n = \Sigma n_i$. Therefore, at a constant absolute temperature T,

$$G_i = -\frac{D_i}{RT} \frac{d(Py_i)}{dx} = -P_i \frac{d(Py_i)}{dx}$$

For nonideal gases and liquids, a compressibility factor must be introduced, as is subsequently shown.

Integrated Form

It may be observed that the gaseous integrated form involves the following mathematical operations, where integration is from the high-pressure or reject side of the membrane to the low-pressure or permeate side:

$$\int d(Py_i) = P_V y_i - P_L x_i$$

where, in the terminology to be adopted,

P_V = pressure of permeate V on the low-pressure permeate side;
P_L = pressure of reject L on the high-pressure reject side;
y_i = composition of permeate V;
x_i = composition of reject L.

That is to say, the permeate stream and its rate are designated by V and the reject stream by L. Integrating across the membrane, between limits,

$$\int dx = \Delta m$$

where Δm is the membrane thickness. Therefore, correcting for the minus sign,

$$G_i = P_i \frac{P_L x_i - P_V y_i}{\Delta m}$$

which is the fundamental membrane pointwise flux equation and can, in turn, be shown to be the same also for liquid permeation and solution permeation. Accordingly, the various relationships that apply for the separation of gases also apply to the separation of liquids and solutions. Note, for a pure gaseous component i, that $x_i = y_i = 1$ such that

$$G_i = P_i \frac{P_L - P_V}{\Delta m}$$

This form is fully equivalent to the mass or molar flux form of the Darcy relation for flow through porous media[3,4,7] (also see Example 1.3); that is,

$$P_i = (K/\mu_i)\rho$$

where, here, K is the medium permeability to fluid flow and μ_i is the fluid viscosity for component i or a mixture i of constant composition, in consistent units. The ratio of the permeability to the viscosity is the mobility, in the units of volumetric flux per unit pressure gradient. By multiplying by the fluid density, whether gas or liquid, the result is the membrane permeability in the corresponding and consistent mass or molar dimensions. That is, multiplication of the mobility by the mass or molar density ρ or ρ_m for the component or mixture gives the mass or molar flux form.

Liquid Systems

Correspondingly, the absolute activity in a liquid phase could be expressed merely as

$$\gamma_i = (vp)_i x_i \quad \text{or} \quad \gamma_i = H_i x_i$$

and the differential rate equation adapted or modified, and integrated accordingly.

However, Fick's law for diffusion is also applicable to diffusion within liquids or by liquids, which can be written

$$G_i = -D_i \frac{dc_i}{dx}$$

where D_i is the diffusion coefficient and c_i is the concentration. We see that the units for D_i are again

$$\frac{\text{moles of } i}{\text{time-area-}\dfrac{\text{moles of } i}{\text{volume-distance}}} = \frac{(\text{distance})^2}{\text{time}}$$

Note that the diffusion coefficient D_i could as well be called the *permeability coefficient*, as pertains to the flow of one substance through, within, or relative to another substance or medium. However, if concentration is the driving force, arguably the coefficient D_i as such can be used, by analogy with Fick's law. Whereas if partial pressure is the driving force, then permeability is more apt for component flow under a partial pressure gradient or difference. In net effect, the symbolism is interchangeable and may incorporate other entities, such as overall membrane thickness to yield an overall permeability coefficient, and the conversion between concentration and partial pressure.

Here, the absolute activity can be assumed to be the concentration c_i. On a molar basis, therefore, the units for D_i are as previously noted:

$$\frac{\text{moles of } i}{\text{time-area-}\dfrac{\text{moles of } i}{\text{volume-distance}}} = \frac{(\text{distance})^2}{\text{time}}$$

which are the customary units for the diffusion coefficient or diffusivity, including interdiffusion, notably the diffusion of gases into gases, and also for diffusion of gases within liquids or solutes within liquids. Alternately, however, for flow through a membrane barrier, the diffusion coefficient can be viewed as a permeability coefficient for component i, say, P_i or P_i^0 or some other symbol. (The notational possibilities are numerous, depending on the system under consideration. This is an argument for utilizing the same permeability symbol, regardless, and define it in the context of its use in the particular rate or flux equation.)

Note in passing that, for the flow of a liquid through a membrane barrier, the permeability behavior is commonly expressed in terms of cc or ml of liquid. Therefore, we may speak of a permeability coefficient P_i^0, whereby the expression for, say, the molar flux

$$G_i = -D_i \frac{dc_i}{dx}$$

becomes

$$\text{volumetric flux of component } i = \frac{G_i}{(\rho_m)_i} = -\frac{D_i}{(\rho_m)_i} \frac{dc_i}{dx} = -P_i^0 \frac{dc_i}{dx}$$

where $(\rho_m)_i$ is the molar density of component i. Accordingly the units for P_i^0 would be

$$\text{time-area} \frac{\dfrac{\text{volume of } i}{\text{moles of } i}}{\text{total volume-distance}}$$

For simplicity, however, let $P_i^0 \equiv P_i$, in the units prescribed, but these must be consistent within the formula employed.

Density vs. Concentration

Incidentally, the molar density $(\rho_m)_i$ of component i is the moles of i divided by the volume of the system and is therefore the molar concentration c_i. Similarly, the mass density ρ_i of a component i is its mass concentration. The basis, of course, is a unit volume, in whatever units are employed. (For a pure component, the volume of component i is the same as the volume of the system. Generalizing, for a mixture, we can speak of the partial or contributing volumes of the components, which add up to the volume of the mixture.)

The component concentration, moreover, can be converted to partial pressure (and mole fraction) via the equation of state for the total liquid phase, utilizing the compressibility factor, as is subsequently shown.

Alternatively, the component density can be converted to pressure via the compressibility form for a (compressible) liquid, also demonstrated subsequently. We therefore note a distinction between what are called *liquid compressibility* and *compressibility factors*.

Solution Systems

For the flow of a dissolved (or suspended) solute species across a membrane, Fick's law can be presumed to hold. That is, the molar flux of the solute species of component i can again be represented in terms of the molar flux, whereby

$$G_i = -D_i \frac{dc_i}{dx}$$

and the units for D_i are again

$$\frac{\text{moles of } i}{\text{time-area-} \dfrac{\text{moles of } i}{\text{volume-distance}}} = \frac{(\text{distance})^2}{\text{time}}$$

and D_i is equivalently a permeability coefficient, representable also as, say, P_i or by some other symbol.

However, the solute concentration can be converted to mole fraction via the equation of state for the liquid phase, as is subsequently shown.

Often, an object of equal interest is the solvent component itself and its permeation. Considered as a pure component or mixture of constant composition, the solvent becomes subject to Darcy's law for flow through porous media. The notable solvent example is water, which most usually pertains to the separation operations of dialysis and reverse osmosis. Beyond this, water generally functions as the particulate-carrier in microfiltration and ultrafiltration.

2.2 PERMEABILITY RELATIONSHIPS AND UNITS

The foregoing derivations have mostly avoided specific assumptions about units. Instead, the equations used may be described as having "consistent" units. For instance, the units of diffusivity or permeability must be consistent with the units of the other terms.

The degree of separation and recovery, however, is independent of the units and depends only on relative permeability values or selectivity, although actual membrane size (area) depends on the particular units used.

Therefore, if the actual numerical values for permeability, pressure, and flow rate are to be used, then specific units are necessarily involved, which in turn, determine membrane area, given the membrane thickness, or the overall membrane permeability.

Pointwise Permeability

For a pure gas, of a single component only, commonly used units for the membrane pointwise permeability P_i for a pure component i are

$$\frac{10^{-9} \, cm^3 \text{(at STP)}}{sec\text{-}cm^2\text{-}cm \ Hg/cm}$$

that is, 10^{-9} cubic centimeters (of gas corrected to standard conditions) per square centimeter per centimeter of mercury pressure change per centimeter. The last-mentioned item in the denominator is the pressure gradient: the total pressure drop divided by the membrane thickness. The units designated centimeters of mercury (Hg) per centimeter therefore pertain to the pressure gradient within the membrane. Alternately, these units are expressible as

$$\frac{10^{-9} \, cm^2}{sec\text{-}cm \ Hg}$$

which, when multiplied by the operating pressure difference in the appropriate and consistent units, becomes the diffusion coefficient or diffusivity; here, it would be measured in 10^{-9} cm^2/sec.

Other units used for the permeability of various solids to gases are presented in the *International Critical Tables*[8] (vol. V, pp. 76–77), as follows:

$$\frac{10^{-6} \, cm^2}{sec\text{-}atm} \quad or \quad \frac{10^{-9} \, cm^2}{sec\text{-}atm}$$

The former units are used for, say, gases through metals and rubber; the latter units for, say, gases through glass. To convert to the pressure units of cm Hg, the numerical values in the these units would be divided by 76 cm Hg/atm. Some representative values for hydrogen taken from the *International Critical Tables* (vol. V) are furnished in Appendix 1. (Contiguous sections in vol. V furnish data about the interdiffusion of gases and vapors, the diffusion of dissolved solids and liquids within liquids, and the diffusion of dissolved solids, ions, and gases within solids.[8]) As noted, hydrogen permeabilities increase with temperature. A side effect of hydrogen is that it can dissolve into the interstices of metals, affecting the ductility and strength, called hydrogen *embrittlement*. Additionally, hydrogen is reactive, notably as pertains to impurities in the metal; that is, components and phases other than the pure metallic state itself. A prime example of the foregoing is with steels and their makeup.

(The second set of units, in the preceding equation, is obviously the smaller collective unit, which would be obtained from the first by a factor of 1000. In other words, a permeability value of, say, 2 in the first set of units would convert as follows:

$$2\frac{10^{-6}\,\text{cm}^2}{\text{sec-atm}} = 2000\frac{10^{-9}\,\text{cm}^2}{\text{sec-atm}}$$

Whichever set of units is preferred depends in part on the magnitude of the membrane permeability itself.)

Henceforth, P_i or, say, P_j is used for the permeability of an arbitrary component i or j; that is, both i and j denote the key components for the separation, especially for a two-component system but also in a multicomponent system. Inasmuch as the symbol P also is used to denote pressure, as is the common practice, some other symbol could be adopted for permeability, say, lower case, script, boldface, Greek, or the like. However, subscripted P_i and P_j seem self-evident, so pressure also is subscripted to provide its distinguishing feature.

Thus, P_L stands for the higher or upstream pressure (or feed-reject pressure) at the membrane, and P_V stands for the lower pressure or downstream pressure (or permeate pressure), with the difference $P_L - P_V$ denoting the pressure-drop across the membrane proper. The analogy is akin to that used for phase separations. Furthermore, the usage, in the main, appears self-evident in context.

The flow is stated in cubic centimeters at standard conditions of temperature and pressure; that is, in standard cubic centimeters. The emphasis here, therefore, is for gases rather than liquids and pertains to a pure component.

The equivalent statement for the flow of liquids through membrane barriers involves the same units. However, the volume is in actual cc or ml of liquid flowing, and of course, other units for the pressure difference may be used. The permeability value ordinarily refers to the pure component and is so measured.

Overall Permeability

Overall permeability for the component of interest can be specified by dividing by the membrane thickness Δm, and the corresponding result would be in the units of

$$\frac{10^{-9}\,\text{cm}^3}{\text{cm}^2\text{-sec-cm Hg}} = \frac{10^{-9}\,\text{cm}}{\text{sec-cm Hg}}$$

where cm Hg corresponds to the total pressure-drop across the membrane in centimeters of mercury. Accordingly, for an assigned or affixed pressure drop, the generalized units for the overall permeability become that of (distance × time^{-1}).

By multiplying by a linear dimension of the membrane, the overall permeability would be measured as

$$\frac{10^{-9}\,cm^3}{cm\text{-sec-cm Hg}}$$

If, instead, multiplied by the total area of the membrane, the overall permeability would be measured as

$$\frac{10^{-9}\,cm^3}{sec\text{-cm Hg}}$$

It is understood that this relationship can be based on the unit cross-sectional area normal to flow; that is, on a flux relationship.

The area of the membrane may be designated as the interfacial area and measured either as the inner or outer surface of the membrane or the mean or average, as in the case of a hollow membrane. A log mean value can also be adapted, in common with the practice for the conduction of heat through a tube wall, as previously observed.

The permeability as measured in the cgs system may also be converted to the English system, even to the units of darcies or millidarcies, for the flow of fluids through porous media, as employed in the petroleum industry.[3,4,7]

The diffusion coefficient, or diffusivity, is used as the measure of the diffusion of one substance through another. The driving force is the concentration difference, via Fick's law, which can be related to the partial pressure difference by means of the equation of state. It is applicable to mixtures occurring as a common phase but can as well be applied to the case where the second substance is a dissolved (or colloidal) solid, say, as in dialysis. The dimensions for the diffusion coefficient are ordinarily (distance)^2/time; for example, cm^2/sec. Since the equation of state is implicitly involved, the conversion between permeability and diffusivity is more pronouncedly temperature and pressure dependent (albeit a constant temperature is generally assumed for membrane performance).

Finally, inasmuch as it is usually more straightforward to work in moles and mole fraction compositions when material balances are involved, it is preferable to put the permeability on a molar basis.

Selectivity

The relative permeability of one component to another is also called the *selectivity*, α, and sometimes the *permselectivity*. In the customary notation used,

$$\alpha = \alpha_{i-j} = P_i / P_j$$

is the permeability of component *i* relative to component *j*. It is related to the degree of separation that may be attained. Some representative values of permeability and selectivity for gases are shown in Appendix 1.

The term *selectivity* is also expressed in the terms of component ratios, which in effect are but the permeability ratios for the pure components. Unfortunately, this does not necessarily mean that the components permeate in the same relative way in mixtures.

Selectivity Factor

Another concept utilized is that of the selectivity factor β_{i-j}, which is the ratio of component concentrations in the permeate divided by the ratio of the component concentrations in the reject or retentate; that is,

$$\beta_{i-j} = \frac{(c_i)_V / (c_j)_V}{(c_i)_L / (c_j)_L}$$

where *c* refers to molar or mass concentration and, in the notation to be used, the subscript *V* stands for the permeate phase and *L* for the reject or retentate phase. In terms of mole or mass fractions of a component *i*, for a binary system,

$$\beta_{i-j} = \frac{\dfrac{y_i}{1 - y_i}}{\dfrac{x_i}{1 - x_i}}$$

where, in the notation to be subsequently used, *y* is the mole or mass fraction of a component in the permeate phase *V* and *x* is the mole or mass fraction of a component in the reject or retentate phase *L*. The preferred units are therefore in terms of mole fraction, this being more compatible with the equation of state and other physical (and chemical) laws for phase behavior; plus, the mole fractions in a distinct phase always sum to unity. However, the literature utilizes various units, which are not always made clear. Representative data are shown in Appendix 1.

Membrane Permeability Units and Terms in Context

In the derivations and examples that follow, the permeability symbol P_i refers to any of the aforementioned classifications of units, but its particular units are to be consistent in the context of usage. For most purposes, but not always, the permeability is an overall or mean permeability.

A few examples are included at the end of the chapter to show various calculations for the conversions between units, as indicated already. Spreadsheet type calculations are shown in Appendix 2.

Membrane Permeabilities in Mixtures

In mixtures, the pressure dimension in the permeability coefficient becomes the partial pressure pp_i. That is, the pointwise permeability coefficient in a mixture will have the units or dimensions of

$$\frac{\text{moles of } i}{\text{time-area-}\dfrac{pp_i}{\text{distance}}}$$

where, as noted previously, $pp_i = Py_i$ by definition. Similarly, the overall permeability coefficient has the dimensions

$$\frac{\text{moles of } i}{\text{time-area-}pp_i}$$

The practice here is not to attempt to adapt symbols to distinguish one kind of permeability coefficient from another (e.g., a pointwise coefficient P_i from an overall coefficient, say $\overline{P}_i$). Rather, as noted earlier, all coefficients are simply denoted P_i, to be further distinguished by context of usage, with the required units or dimensions and magnitude dictated by that usage.

Conversion of Liquid-Phase Permeation to Gas-Phase Format

The driving force for liquid phase permeation is regarded as a concentration difference (via Fick's law) rather than a partial pressure difference. Viewed in terms of absolute activity γ as the driving force (i.e., as a potential function), the absolute activity for a component i in the liquid phase is in general different than in the vapor phase (albeit at a vapor-liquid equilibrium condition, they must be the same). That is, the nonequilibrium absolute activity in the vapor or gas phase is the partial pressure, whereas in the liquid phase it is judged to be the concentration.

There are ramifications: The driving force in the liquid phase is mostly pressure independent, depending mainly on concentrations, whether speaking of a miscible liquid component or a dissolved solute in a solution. This is reported to be borne out experimentally but not conclusively.[9(pp. 279–281)]

The object, therefore, is to reconcile concentration-driven liquid-phase or solution-phase permeation with partial pressure–driven gas-phase permeation, in that the extensive relationships developed subsequently for the latter can be conveniently used for the former. That is, gas permeation relationships are to constitute the more convenient, baseline circumstance but are fully amenable to calculating liquid phase behavior in terms of mole fractions as well.

Fundamentally, it becomes the transformation from concentration units to mole fraction units. And, whereas the permeation of a pure liquid may be expressed for instance in terms of its liquid volume per se, the permeation of a component (or components) from a liquid mixture requires the introduction of the idea of composition; namely, the component concentration or mole fraction. For the purposes here, the mole fraction is regarded as of more utility than component concentration. (For one thing, the totality of component mole fractions always sums to unity.) Therefore, the conversion is from concentration units to mole fraction units.

The starting point for developing the equivalence is the nonideal gas law, or equation of state, which more generally can be made applicable to any single phase or single-phase region, whether represented as gas or liquid. This relationship is representable as

$$PV = znRT$$

where the customary units are

P = pressure;
V = volume of the system;
n = the total number of moles present;
R = the gas constant in pressure-volume units;
z = the compressibility factor, a measure of nonideality, which for an ideal gas is unity.

The compressibility factor z is, in turn, a function of pressure and temperature and, for a pure component, depends also on the identity of the particular component. For a mixture, z in general depends on the identity of the components and their compositions. Generalized correlations are available, in which behavior of the compressibility factor is graphically represented in terms of the reduced pressure and temperature.[10] For a pure component, the reduced value is the actual pressure or temperature divided by the critical pressure P_c and critical temperature T_c; that is, $P_r = P/P_c$ and $T_r = T/T_c$. For mixtures, the pseudocriticals P_{pc} and T_{pc} are used, which are the sums of the mole fractions times the respective criticals, such that $P_r = P/P_{pc}$ and $T_r = T/T_{pc}$.

It may be observed that the molar density ρ_m is given by

$$\rho_m = \frac{n}{V} = \frac{P}{zRT}$$

whereas the mass density ρ is given by

$$\rho = \rho_m(MW) = \frac{(MW)P}{zRT}$$

where MW is used to denote molecular weight.

By virtue of Dalton's law (or definition) for mixtures, for a particular component i, it may be written that its partial pressure pp_i is given by

$$pp_i = Py_i = P\frac{n_i}{n} = zn_i\frac{RT}{V}$$

where y_i is the mole fraction of component i present in the mixture and n_i is the number of moles of component i present. For convenience, the behavior of z can be assumed to remain the same as that for the total mixture. Furthermore, as indicated,

$$n = \sum n_i \qquad \text{and} \qquad y_i = \frac{n_i}{n}$$

More generally speaking, these equation forms may be applied to liquids as well as gases; that is, to any single-phase fluid, whether called a gas or a liquid. (Moreover, the mole fraction symbol y_i can as well be replaced by x_i, the latter most usually pertaining to what is thought of as a liquid. However, in the nomenclature to be used in subsequent chapters, the symbol y is reserved for the permeate mole fractions and x for the reject.)

Strictly speaking, the terms gas and liquid pertain only to an equilibrium between the two phases. The less-dense phase is the gas or vapor, the more-dense phase is the liquid. Moreover, it is possible to go from one to the other in P-T space by circumventing the critical point, the point reached in P-T space at which the vapor and liquid can no longer coexist at saturation.

Hence, the terminology used may be for a single-phase region, which is roughly compartmentalized into a superheated vapor region (less-dense single-phase region) and a supercooled liquid region (more-dense single-phase region). The region above and beyond the critical point is called the *supercritical region*.

Note, in the foregoing, that the ratio n_i/V in the expression for partial pressure is also the concentration c_i, which is also equal to the molar density:

$$c_i = \frac{n_i}{V} = \frac{Py_i}{zRT}$$

This provides the connection between partial pressure (that is, pressure times mole fraction) and concentration. Therefore, the following substitution can be made for the molar flux:

$$G_i = -D_i \frac{dc_i}{dx} = -D_i \frac{1}{zRT} \frac{d(Py_i)}{dx} = -P_i \frac{d(Py_i)}{dx}$$

where the permeability P_i used here is defined by the substitution. Note that this equation is of the same form as for gaseous permeation and that the symbol y_i is used for the component mole fraction in the liquid phase, albeit x_i could be used instead.

Accordingly, the relationships derived for gaseous permeation can be used for liquid permeation by suitably modifying the permeability coefficient; that is, for liquid permeation, let

$$P_i = D_i \frac{1}{zRT}$$

Whereas the gas constant R and the absolute temperature T are known or given, the permeability conversion requires the determination of the behavior of the compressibility factor z for a liquid.

Fortuitously, this kind of information is available, as per Part II of Hougen, Watson, and Ragatz's Chemical Process Principles.[10] Figure 137[10(p. 574)] shows a correlation for the compressibility factor for a saturated liquid in terms of reduced pressure. The value varies from near 0 up to about 0.3 at the critical point, where the reduced pressure becomes unity. In Figure 140,[10(p. 580)] the logarithmic behavior of the compressibility factor is given for both gases and liquids, in the one case with reduced pressure as the abscissa and in the other case with reduced temperature as the abscissa.

Since the behavior of a supercooled liquid can be assumed not to vary too appreciably from its behavior at saturation or phase equilibrium, the former correlation ordinarily suffices.

For an overall or integrated change across the membrane between the liquid reject phase and the liquid permeate phase, it follows that, in consistent units,

$$G_i = P_i \frac{P_L x_i - P_V y_i}{\Delta m}$$

However, since the compressibility factor z also varies during the integration, a mean value can be used in the permeability conversion. Thus,

$$P_i = D_i \frac{1}{\bar{z}RT}$$

where $\bar{z}$ is the mean compressibility factor. Since temperature is regarded as constant, an estimate can be made from the Hougen, Watson, and Ragatz correlations for z against reduced pressure (P_r) for a saturated liquid. In logarithmic or exponential form, this correlation is approximately represented by

$$z = 0.161(P_r)^{0.978} \qquad \text{or} \qquad z \sim 0.17(P_r)$$

which can be used to estimate a value for $\bar{z}$ from the (averaged) reject and permeate pressures. These expressions are of the same form as for gas-phase permeation.

Liquid-Phase Density

Note that the mass density ρ of a vapor or liquid phase can be represented in term of the compressibility factor as

$$\rho = \frac{MP}{zRT}$$

where M denotes the molecular weight $(M \equiv MW)$. If the compressibility behavior can be represented by the relationship

$$z = a(P_r)^n = \frac{a}{(P_{pc})^n} P^n$$

then it follows that

$$\rho = \frac{MP}{\dfrac{a}{(P_{pc})^n} P^n} = \left[\frac{M(P_{pc})^n}{a}\right] P^{1-n}$$

where the term in brackets can be regarded as a constant.

If $n = 1$, then the liquid phase can be viewed as an incompressible fluid. However, if, say, $n = 0.978$ as per the more rigorous curve fit, then a degree of compressibility can be said to exist.

By referencing the density to a standard density ρ_0 at pressure P_0, it follows that

$$\frac{\rho}{\rho_0} = \left(\frac{P}{P_0}\right)^{1-n}$$

In fact, this expression for density may be compared with the form for liquid compressibility, as is subsequently shown.

Conversion of Solution Permeation to Gas-Phase Format

The conversion is essentially as performed for liquid-phase permeation, save that only the volume term is involved; that is,

$$c_i = \frac{n_i}{V} = \frac{n_i}{\dfrac{znRT}{P}} = \frac{Py_i}{zRT}$$

where y_i is the mole fraction of solute. The flux equation again becomes

$$G_i = -D_i \frac{dc_i}{dx} = -D_i \frac{1}{zRT} \frac{d(Py_i)}{dx} = -P_i \frac{d(Py_i)}{dx}$$

whereby

$$P_i = D_i \frac{1}{zRT}$$

For an overall change, as for liquids,

$$G_i = P_i \frac{P_L x_i - P_V y_i}{\Delta m}$$

$$P_i = D_i \frac{1}{\bar{z}RT}$$

where $\bar{z}$ is the mean compressibility factor. Again, as an approximation for z,

$$z = 0.161(P_r)^{0.978} \qquad \text{or} \qquad z \sim 0.17(P_r)$$

which can be used to estimate a mean value for $\bar{z}$. Again, these expressions are the same as for both gas-phase permeation and its liquid-phase counterpart.

Pressure-Independent Form for Liquids and Solutions

It may be observed that the previously presented basic expression for Fick's law does not include pressure in either the differential or integrated form:

$$G_i = -D_i \frac{dc_i}{dx} = -D_i \frac{\Delta c_i}{\Delta m}$$

or

$$\text{volumetric flux} = \frac{G_i}{\rho_m} = -\frac{D_i}{\rho_m} \frac{dc_i}{dx} = -P_i^0 \frac{dc_i}{dx} = -P_i^0 \frac{\Delta c_i}{\Delta m}$$

where ρ_m is the molar density, assumably a near constant. Accordingly, the flux or permeation rate may be largely independent of the pressure difference, as indicated later in Table 2.1, albeit a pressure difference is ordinarily considered vital to permeation. The experiments for Table 2.1 pertain to pervaporation, which involves a phase change and is a composite of liquid permeation followed by vapor permeation. Incidentally, no material balances are provided; that is, for the closure of the feed utilized with the reject and permeate produced. Therefore, note that the permeation data for organic liquids has been questioned.[9(pp. 278–279)]

Effect of Compressibility Factor

The discrepancy from the foregoing gas-phase equivalent expressions occurs in the introduction of the compressibility factor z via the equation of state, the behavior of which is accommodated by assuming a mean value. More properly speaking, the integration between limits in the mass or molar flux form should appear as

$$G_i = -D_i \frac{dc_i}{dx} = -D_i \frac{\Delta c_i}{\Delta m} = -D_i \frac{\left(\dfrac{n_i}{n}\dfrac{P}{zRT}\right)_{\text{permeate}} - \left(\dfrac{n_i}{n}\dfrac{P}{zRT}\right)_{\text{reject}}}{\Delta m}$$

or, in the previously assigned notation for the reject and permeate phases and multiplying through by the negative sign,

$$G_i = -D_i \frac{dc_i}{dx} = -D_i \frac{\Delta c_i}{\Delta m} = D_i \frac{\dfrac{P_L x_i}{z_{\text{reject}}RT} - \dfrac{P_V y_i}{z_{\text{permeate}}RT}}{\Delta m}$$

whereby the behavior of z can in part tend to cancel out the behavior of the pressure. That is, for liquids, the compressibility factor z may be perceived as increasing almost linearly with P, and the ratio P/z therefore tends to be constant but not absolutely so.

Reduced Pressure

In terms of the previous data fit for the reduced pressure, making the substitution that $z = aP_r = a(P/P_{pc})$, where $a \sim 0.17$, we obtain

$$G_i = D_i \frac{\dfrac{P_L x_i}{z_{\text{reject}}RT} - \dfrac{P_V y_i}{z_{\text{permeate}}RT}}{\Delta m} = \frac{D_i}{aRT} \frac{(P_{pc})_{\text{reject}} x_i - (P_{pc})_{\text{permeate}} y_i}{\Delta m}$$

$$= P_i \frac{(P_{pc})_{\text{reject}} x_i - (P_{pc})_{\text{permeate}} y_i}{\Delta m}$$

Accordingly, the gas-phase partial-pressure format could be used by substituting

$$P_L = (P_{pc})_{\text{reject}} \qquad \text{and} \qquad P_V = (P_{pc})_{\text{permeate}}$$

and adapting the permeability coefficient to incorporate a as indicated.

If the pseudocriticals for the reject and permeate are approximately the same or a mean value is utilized, say, $\sim\bar{P}_{pc}$, then the preceding reduces to

$$G_i = \frac{D_i(\bar{P}_{pc})}{aRT} \frac{x_i - y_i}{\Delta m} = P_i \frac{x_i - y_i}{\Delta m}$$

with the permeability coefficient adapted or modified as indicated.

Therefore, the flux depends more or less on a mole fraction difference. Therefore, the preceding form could be used in the gas-phase calculations if $P_L = P_V = 1$, where the permeability coefficient incorporates $(\bar{P}_{pc})$. Otherwise, $P_L = P_V = (\bar{P}_{pc})$ for the averaged mixture.

The membrane thickness Δm may, of course, be incorporated into the pointwise diffusion coefficient D_i or the equivalent permeability coefficient P_i to obtain the overall coefficient.

Pure Liquids or Liquid Mixtures of Constant Composition

A problem in the preceding representation occurs whenever the flux depends mainly on a mole fraction difference for the limiting or boundary condition for a pure component or a mixture of constant composition. Under this circumstance, the pseudocritical pressure remains the same. Furthermore, the flow rate or flux would be required to be zero and is not directly proportional to the pressure difference, as required for the flow of fluids through porous media. It may be assumed, therefore, that this flux relationship in terms of pseudocritical pressures is not an allowable representation.

For the special case of a pure liquid or a liquid mixture of constant composition, the mole fractions remain equal, such that $x_i = y_i$, and the pressure-independent form equates to 0. (Similarly, for a mixture of constant composition, $x_i = y_i$.) This dilemma is avoided by assuming that the compressibility factor z remains the same for both reject and permeate (which, strictly speaking, is not the case, since the pressure varies) or by introducing a mean value. Then, for the mass or molar flux in consistent units,

$$G_i = -D_i \frac{dc_i}{dx} = -D_i \frac{\Delta c_i}{\Delta m} = D_i \frac{1}{\bar{z}RT} \frac{P_L - P_V}{\Delta m} = P_i \frac{P_L - P_V}{\Delta m}$$

This result is fully equivalent to the Darcy relation for the flow of fluids through porous media, where, dropping subscripts for the permeability P, where $P = (K/\mu)\rho$, the relationship is the same as previously established for a pure gas or a mixture of constant composition (also see Example 1.3). The ratio of the flow permeability K of the medium to the fluid viscosity μ becomes the mobility K/μ. Furthermore, the product $(K/\mu)\rho$ is equivalent to the membrane permeability in terms of mass flux, and $(K/\mu_i)\rho_m$ is equivalent to the membrane permeability in terms of molar flux. This relationship applies to the permeation of pure components only or to the permeation of a mixture of constant composition.

Liquid Compressibility

Another option for pure liquids or mixtures of constant composition is to adapt the commonly used formula for the density ρ of a slightly compressible liquid in terms of its (liquid) compressibility c.[7(p. 42)] The density relationship follows:

$$\rho = \rho_0\, e^{c(P-P_0)} = [\rho_0\, e^{-cP_0}]e^{cP} = A\,e^{cP}$$

where ρ_0 and P_0 denote the density and pressure at a reference condition and the constant A is defined by the substitution. Alternately, for the further purposes here,

$$\frac{\rho}{\rho_0} = e^{c(P-P_0)}$$

in terms of the reference quantities. Furthermore, the temperature is considered to remain at a constant value.

The foregoing is based on the pointwise expression for the coefficient of volumetric expansion at constant temperature, customarily denoted by β:

$$\beta = -\frac{1}{V}\left(\frac{\partial V}{\partial P}\right)_T$$

where the volume V may be based on unit mass or unit mole. For liquids, the values of β are generally small, circa 10 to 100×10^{-6} atm^{-1}, with the notation that β at first decreases rapidly as the pressure increases, then decreases more slowly as the pressure further increases. For example, handbook values for ethyl alcohol are 100 megabars^{-1} at 23 megabars and 14°C and 63 megabars^{-1} at 500 megabars and 20°C (where 1 bar = 0.987 atmospheres). For water, the values are 49 megabars^{-1} at 13 megabars and 20°C and 43 megabars^{-1} at 200 megabars and 20°C.

If the preceding relationship is integrated between limits at constant temperature, the result can be written as

$$\beta(P - P_0) = \ln\frac{V_0}{V} = \ln\frac{\rho}{\rho_0}$$

whereby the usual mean or integrated expression is attained, so that

$$\frac{\rho}{\rho_0} = e^{\beta(P-P_0)} = e^{c(P-P_0)} = \frac{A}{\rho_0}e^{-cP}$$

where the convention is adapted that, for liquids, $\beta \equiv c$. (Furthermore, and more properly, mean values should be used for the compressibility; that is, $\bar{\beta}$ and $\bar{c}$.)

In turn, the molar density can be represented as

$$(\rho_m)_i = (MW)_i A e^{cP}$$

where $(MW)_i$ or M_i is the molecular weight of a pure component i or a mixture i. As noted, the molar density of a pure component i is the same as its concentration c_i.

It may be further noted that a first approximation for the exponential gives the following reduction in terms of the introduced correlation constant A:[7(pp. 91–92)]

$$\rho \sim A[1 + cP)] = A + AcP$$

where ρ can be assumed approximately linear with P.

Accordingly, the integrated form for Fick's law can be written as

$$G_i = D_i\frac{(c_i)_{\text{reject}} - (c_i)_{\text{permeate}}}{\Delta m} = D_i\frac{[(\rho_m)_i]_{\text{reject}} - [(\rho_m)_i]_{\text{permeate}}}{\Delta m}$$

$$= D_i(MW)_i Ac\frac{P_{\text{reject}} - P_{\text{permeate}}}{\Delta m}$$

This, interestingly, becomes the molar flux form of Darcy's law for the steady-state flow of fluids through porous media.[7(p. 94)] The qualifications for Darcy's law are, however, that the fluid be a pure component or a mixture of a constant composition; that is, relative or selective component diffusion does not occur. It may be added that the constant A, as used here and as defined previously, has the dimensions of density. This relation satisfies the boundary condition that for a pure component or a mixture of constant composition, Fick's law for diffusion translates to the Darcy form for the flow of fluids through porous media.

Compressibility of an Ideal Gas

For an ideal gas, if the coefficient of volumetric expansion β is again denoted as

$$\beta = -\frac{1}{V}\left(\frac{\partial V}{\partial P}\right)_T$$

then substitution of the ideal gas law $PV = RT$ yields

$$\beta = -\frac{P}{RT}\left(-\frac{RT}{P^2}\right) = \frac{1}{P}$$

and β varies inversely with the pressure; this is also noted for liquids, if not linearly or directly, at least in substance.

It should be emphasized that this coefficient of volumetric expansion at constant temperature is different than both the coefficient of volumetric expansion at constant pressure and the coefficient of pressure expansion at constant volume.

For the record, a coefficient of volumetric expansion at constant pressure may be portrayed in the partial derivative form:

$$\alpha^* = \left(\frac{\partial V}{\partial \theta}\right)_P$$

where the symbol θ designates a temperature scale. The partial derivative integrates to

$$V - V_0 = \alpha^*(\theta - \theta_0) \qquad \text{or} \qquad V = V_0[1 + \alpha(\theta - \theta_0)]$$

where $\alpha = \alpha^*/V_0$ and which may be more properly designated as the coefficient of expansion for a gas at constant pressure.

A coefficient for pressure expansion at constant volume may be portrayed by

$$\beta^* = \left(\frac{\partial P}{\partial \theta}\right)_V$$

which integrates to

$$P - P_0 = \beta^*(\theta - \theta_0) \qquad \text{or} \qquad P = P_0[1 + \beta(\theta - \theta_0)]$$

where $\beta = \beta^*/P_0$ and which may be more properly designated as the coefficient of pressure expansion for a gas at constant volume.

Both coefficients α and β are utilized in gas thermometry and, in fact, form the basis for the perfect gas law.[11(pp. 62ff.)] In the centigrade or Celsius scale, it is found that, at more ideal conditions of lower pressure (and moderate temperatures),

$$\alpha \sim \frac{1}{273} \quad \text{and} \quad \beta \sim \frac{1}{273}$$

Based on this observation, it follows that, at constant pressure and constant volume,

$$\frac{V}{V_0} = \frac{273 + (t-0)}{273} \quad \text{and} \quad \frac{P}{P_0} = \frac{273 + (t-0)}{273}$$

where t denotes the temperature in the centigrade or Celsius scale and the reference condition is at 0°C. Therefore, a new temperature scale T is indicated where

$$T = 273 + t \quad \text{and} \quad T_0 = 273 + 0$$

The further manipulations to yield the perfect gas law are shown in the reference. For instance,

$$\left(\frac{\partial T}{\partial V}\right)_P = \frac{T_0}{V_0} = \frac{T}{V} \quad \text{and} \quad \left(\frac{\partial T}{\partial P}\right)_V = \frac{T_0}{P_0} = \frac{T}{P}$$

such that

$$dT = \left(\frac{\partial T}{\partial V}\right)_P dV + \left(\frac{\partial T}{\partial P}\right)_V dP = \frac{T}{V}dV + \frac{T}{P}dP$$

where the latter perfect differential rearranges to

$$\frac{dT}{T} = \frac{dV}{V} + \frac{dP}{P}$$

which integrates to

$$\ln T + \ln R = \ln V + \ln P \quad \text{or} \quad PV = RT$$

where $\ln R$ (or R) is introduced as the constant of integration.

Since β is customarily used to denote both the coefficient of pressure expansion for a gas and the coefficient of volumetric expansion for a liquid, here the latter usage is replaced by the symbol c.

Relation of Liquid Compressibility to Compressibility Factor Behavior

The aforementioned integrated expressions for the density ratio for liquid compressibility and compressibility factor behavior may be equated as follows:

$$\frac{\rho}{\rho_0} = e^{c(P-P_0)} = \left(\frac{P}{P_0}\right)^{1-n}$$

Taking the natural logarithm,

$$c(P - P_0) = (1 - n)\ln(P/P_0)$$

it may be noted that it is required that $1 > n$, as previously developed from a curve fit for the compressibility factor for liquids (where $n = 0.978$). The inference is that either c or n (or both) must be a function of pressure.

Taking the first term of the restricted logarithmic expansion, however,

$$\ln(P/P_0) \sim (P/P_0) - 1 \qquad \text{for } 0 < (P/P_0) \le 2$$

yields

$$c(P - P_0) \sim (1 - n)\frac{P - P_0}{P_0}$$

where

$$c \sim \frac{1 - n}{P_0}$$

By this particular route, the liquid compressibility c can be calculated from n and P_0, the pressure at the standard reference condition. Otherwise, c would be a function of pressure and inversely so, as it is, anyway, based on the experimental evidence in the handbooks.

Since the standard reference condition can be assumed arbitrary, this relationship would constitute a means for estimating a standard reference pressure P_0 from known values of c and n. In other words, nothing has been gained.

Merely as an exercise, if as previously estimated by a curve fit, $1 - n = 1 - 0.978 = 0.022$ and $c \sim 100(10^{-6})$ atm^{-1}, then $P_0 \sim 220$ atm. Pointwise behavior is more revealing. If

$$c = \beta = -\frac{1}{V}\left(\frac{\partial V}{\partial P}\right)_T$$

then for a mole of nonideal gas, at constant temperature, where $PV = zRT$,

$$c = -\frac{1}{zRT}\frac{d\left(\dfrac{zRT}{P}\right)}{dP} = -\frac{1}{z/P}\frac{d(z/P)}{dP} = -\frac{d\ln(z/P)}{dP}$$

or

$$c = -\frac{1}{z/P}\left(\frac{1}{P}\frac{dz}{dP} - \frac{z}{P^2}\right) = -\frac{1}{z}\frac{dz}{dP} + \frac{1}{P} = -\frac{d\ln z}{dP} + \frac{1}{P}$$

Extending the representation to a single, more-dense phase (what we think of as a compressible liquid), if the compressibility factor at constant temperature can be represented by

$$z = aP^n \qquad \text{or} \qquad \ln z = n\ln P + \ln a$$

then

$$c = -\frac{n}{P} + \frac{1}{P} = \frac{1-n}{P}$$

Thus, the pointwise value for the liquid compressibility would be inversely proportional to the pressure and zero for an incompressible fluid (where $n = 1$). Furthermore, for a perfect gas, $n = 0$.

Adaptation to Gas-Phase Format

The integrated relationship for gas-phase permeation has been previously given as

$$G_i = P_i \frac{P_L x_i - P_V y_i}{\Delta m}$$

And, as has been shown, the integrated relationship for the liquid-phase permeation of mixtures can be presented as

$$G_i = \frac{D_i}{RT} \frac{\dfrac{P_L x_i}{z_{\text{reject}}} - \dfrac{P_V y_i}{z_{\text{permeate}}}}{\Delta m} = \frac{D_i}{\bar{z}RT} \frac{P_L x_i - P_V y_i}{\Delta m} = P_i \frac{P_L x_i - P_V y_i}{\Delta m}$$

where z is pressure dependent at constant temperature. The substituted permeability coefficient P_i is in the appropriate corresponding and consistent units:

$$\frac{\text{moles of } i\text{-distance}}{\text{time-area-}\dfrac{\text{moles of } i}{\text{mole}}}$$

This permeation relationship can be used in the separation calculations for liquid mixtures by assuming that $P_i = D_i/\bar{z}RT$ in the gas-phase format and all units are consistent, with the permeability incorporating the averaged or mean compressibility factor for the reject and permeate. Using the mean compressibility factor partially offsets the effect of pressure difference on the flux relationship for component i. In other words, the permeability coefficient itself can be perceived as dependent on the initial pressure P_L and final pressure P_V and changes with the particular operating conditions.

Furthermore, the boundary condition is satisfied that, for a pure liquid or liquid of constant composition, the flow or flux is directly proportional to a pressure difference. That is, Darcy's law for the flow of fluids through porous media is accommodated. However, the permeability value per se changes with each particular circumstance.

Membrane Areas for Mixtures

Interestingly, membrane permeabilities are generally measured and specified for each pure component as such, even though the permeabilities for each component in a mixture may have been found to be less than for the respective pure components, as annotated by Lee and Minhas.[12]

(The analogy is with deviations from Raoult's law for vapor-liquid equilibria, where ideally the vapor-phase partial pressure for each component of a mixture is equal to the vapor pressure of the pure component times its liquid-phase mole fraction. In further explanation, the vapor-phase partial pressure is equal to the total system pressure times the component mole fraction in the vapor phase. In practice, however, usage requires the introduction of an "activity coefficient" for each component to accommodate this departure from ideality. Furthermore, and strictly speaking, the behavior of these activity coefficients varies with temperature, pressure, composition, and the particular components making up the mixture. The most orderly representation is with mixtures of the lighter hydrocarbons, say, methane through heptanes, in terms of the so-called K-values or equilibrium vaporization constants or ratios—read coefficients. This representation is most orderly well away from the critical region for the mixture.)

Consider, therefore, a mixed permeate phase V, which also signifies the molar flow rate. If the mole fraction of a component i present in the permeate phase is designated y_i and the mole fraction present in the feed-reject phase L is denoted by x_i, then the permeation relationship should presumably be of the basic form

$$Vy_i = P_i(P_L x_i - P_V y_i)A$$

where A is the membrane interfacial area and $(P_L - P_V)$ would be the total pressure difference across the membrane proper. However, the partial pressure difference $(P_L x_i - P_V y_i)$ represents the driving force for component i, so to speak, albeit it may be represented otherwise using the idea of an "activity" difference.

Moreover, the units are to be consistent; that is, here the permeability P_i has the units of moles per unit time per unit area per unit pressure (or partial pressure) difference but refers to the permeability measured for the pure component i. As such, it denotes an overall permeability; that is, the membrane thickness has already been taken into account.

The preceding is the basic form adapted for Chapter 3 and the following chapters, albeit the membrane area A may be incorporated into the permeability term for simplification purposes.

Assuming, however, that the permeability is represented in the units of moles/area-time-pressure difference, the membrane area is calculated from

$$A = \frac{Vy_i}{P_i(P_L x_i - P_V y_i)}$$

Again, the units are to be consistent. For the permeation of a pure component only, this yields the expected relationship, where y_i and x_i are unity.

Finally, we emphasize that the subsequent membrane separation derivations and calculations involving two or more components should be—and are—internally consistent. That is to say, for the purposes here, the same membrane area requirement results whether we are dealing with component i, component j, or any other component of the feed mixture. In other words, the equation derivations and calculations are to be simultaneous for each component.

Spreadsheet-type calculations corresponding to the following examples are presented in Appendix 2.

EXAMPLE 2.1

A membrane has a nominal pointwise permeability of 20 in the standard or customary units of 10^{-9} cc^3(STP)/sec-cm^2-cm Hg/cm, as previously set forth, and a thickness of 10 μ or 10 microns, 10×10^{-6} m, or 10×10^{-3} mm, or $10 \times 10^{-4} = 10^{-5}$ cm. A perhaps more useful conversion for the overall permeability is as follows, as per Example 1.1:

$$P_i = \frac{20(10^{-9})}{10(10^{-4})}\frac{76}{22,414} = 20(0.00339)(10^{-6})\frac{\text{g-moles of } i}{\text{cm}^2\text{-sec-atm}}$$

where 22,414 is standard cc/g-mole at 1 atm pressure and 0°C (and 76/22,414 = 0.00339). Alternately,

$$P_i = 20(0.00339)(10^{-6})(30.48)^2\frac{1}{453.59}\frac{1}{14.696}$$

$$= 20(0.000472)(10^{-6})\frac{\text{lb-moles of } i}{\text{ft}^2\text{-sec-psi}}$$

$$= 20(1.700)(10^{-6})\frac{\text{lb-moles of } i}{\text{ft}^2\text{-hr-psi}}$$

$$= 34(10^{-6})\frac{\text{lb-moles of } i}{\text{ft}^2\text{-hr-psi}}$$

Hence, there is a choice of values and units, depending on the circumstance.

EXAMPLE 2.2

Apropos of Example 2.1, a membrane cell is to have the following pointwise permeabilities to components i and j:

$$P_i = 20(10^{-9})\ \text{cm}^3/\text{cm}^2\text{-sec-cm Hg/cm}$$
$$P_j = 10(10^{-9})\ \text{cm}^3/\text{cm}^2\text{-sec-cm Hg/cm}$$

The pressure P_L on the high-pressure or reject side of the membrane and the pressure P_V on the low-pressure or permeate side of the membrane are as follows:

$$P_L = 3(10^1) \text{ atm} \quad \text{or} \quad 30 \text{ atm}$$
$$P_V = 2(10^1) \text{ atm} \quad \text{or} \quad 20 \text{ atm}$$

For a membrane thickness of 10 μ (10 microns or 10×10^{-4} cm), the pointwise permeabilities convert to overall permeabilities of

$$P_i = 20(76/22,414)(10^{-6}) \text{ g-moles/cm}^2\text{-sec-atm}$$
$$P_j = 10(76/22,414)(10^{-6}) \text{ g-moles/cm}^2\text{-sec-atm}$$

where again the number 22,141 is the standard in cc/g-mole (at 1 atm pressure and 0°C and where 76/22,414 = 0.00339). Therefore, the product of the permeabilities in these dimensions times the pressure or pressure difference in atm yields a corresponding flux value in g-moles/cm²-sec:

$$G_i = 20(76/22,414)(10^{-6})(30-20) = 0.678(10^{-6}) \text{ g-moles of } i/cm^2\text{-sec}$$
$$G_j = 10(76/22,414)(10^{-6})(30-20) = 0.339(10^{-6}) \text{ g-moles of } j/cm^2\text{-sec}$$

Anticipating the problem statement and results to be used in Example 3.1 of Chapter 3, for a single-stage separation, the overall permeabilities and the pressures may be assigned as yet arbitrary or unspecified dimensions such that

$$P_i = 20 \quad P_L = 3$$
$$P_j = 10 \quad P_V = 2$$

In other words, the units or dimensions may be assigned after the calculations rather than before. Furthermore, the arbitrary or relative values are all that are needed to determine the degree of separation attainable, as demonstrated in Example 3.1. The absolute values are needed only for determining the membrane area, the object of this exercise.

As per Example 3.1, for the arbitrary characteristics so listed, an arbitrary permeate rate is determined with the value $V'' = 12.7056$ for an assigned permeate to feed (V/F) ratio of 0.5 and, for component i, has the dimensions of overall permeability times pressure $(P_i P_V)$; that is, it will be in the dimensions of permeate flux, as per the following derived equation from Chapter 3.

To continue, consider the expression for the dimensionless K-value or permeate-reject composition distribution function as derived in Chapter 3,

where

$$K_i = \frac{P_i P_L}{V'' + P_i P_V}$$

As in Example 1.2, the appropriate dimensions can be introduced as a multiplier into both the numerator and denominator:

$$K_i = \frac{[20(3)](10^{-9})\dfrac{1}{10(10^{-4})}\dfrac{76}{22,414}(10)}{V''(10^{-9})\dfrac{1}{10(10^{-4})}\dfrac{76}{22,414}(10) + [20(2)](10^{-9})\dfrac{1}{10(10^{-4})}\dfrac{76}{22,414}(10)}$$

$$= 1.138399$$

where $V'' = 12.7056$, as per the calculations of Example 3.1, and $[20(3)]$ and $[20(2)]$ are the original values of $P_i P_L$ and $P_i P_V$ in terms of the arbitrary dimensions, as used in Example 3.1.

To continue, on introducing the dimensions, the value of the total permeate flux becomes as in Example 1.2:

$$(12.7056)(76/22,414)(10^{-9})/(10^{-4}) = 0.4308(10^{-6}) \text{ g-moles/cm}^2\text{-sec}$$

If the feed rate is, say, 1 g-mole/sec and the permeate to feed ratio is 0.5, then the membrane area requirement again becomes

$$\text{Area} = \frac{1(0.5)}{0.4308(10^{-6})} = 1.16(10^6) \text{ cm}^2$$

per gram-mole of feed/sec. Similar conversions can be made to other units.

EXAMPLE 2.3

The units for membrane permeability behavior in liquid systems may be converted to the gaseous phase format as follows. A distinction must be made, however, for whether the system is a mixture or a pure component. Furthermore, every situation is liable to be different, depending on the kind of information supplied and the particular units involved.

Mixtures

Consider the previously derived formula,

$$G_i = \frac{D_i}{\bar{z}RT}\frac{P_L x_i - P_V y_i}{\Delta m} = P_i \frac{P_L x_i - P_V y_i}{\Delta m}$$

where D_i is the liquid phase pointwise permeability coefficient or diffusion coefficient for component i in concentration units. Furthermore, the compressibility factor z may be approximated by $z = aP$ upward toward the critical point, where a has the value of ~0.17, as was obtained from a data fit for the compressibility factor of liquids in terms of the reduced pressure (assuming constant temperature). For the record, for a mixture, the pseudo-critical is the summation of the criticals times the corresponding mole fractions. This equation is compatible with the gas phase format, but requires that a mean or average value be introduced for the compressibility factor.

As an illustration of the conversion calculations involved, consider the information contained in Tables 2.1 and 2.2 for the n-heptane-isooctane system using a membrane of 1 mil thickness, for operating conditions of 100°C or 212°F (with liquid feed and vaporized permeate, constituting the composite behavior called *pervaporation*). The permeate flux was apparently found to be independent of the feed-reject pressure. This may or may not be contradictive.

Some properties for the individual components n-heptane and isooctane are tabulated and compared as in Tables 2.1 and 2.2 (which obviously form closely boiling mixtures). The feed-reject inlet pressure is 15 psig or 30 psia in one case and 115 psig or 130 psia in the other.

Some useful and appropriate determinations for the feed-reject and permeate are as follows, albeit no material balances (and their closure)

Table 2.1 n-Heptane and Isooctane Properties

Component	Specific Gravity	P_c (in psia)	MW	Boiling Point at 1 atm °F	Boiling Point at 1 atm °C	Vapor Pressure at 100°F (37.78°C)
n-Heptane	0.6883	396.8	100.2	209.16	98.42	1.6201 psia
Isooctane	0.6962	372.5	114.2	210.63	99.24	1.7089

Table 2.2 Pressure-Independence for Liquid-Phase Permeability and Selectivity

Feed Pressure	15 psig	115 psig
Operating temperature	100°C	100°C
Feed composition	50 vol % n-heptane 50 vol % isooctane	50 vol % n-heptane 50 vol % isooctane
Permeate composition	75 vol % n-heptane 25 vol % isooctane	75 vol % n-heptane 25 vol % isooctane
Permeate flux	140 gal/sq ft-hr × 10^3	140 gal/sq ft-hr × 10^3

Source: Adapted from Kesting[9(p. 281)] based on data from R. C. Binning, R. Lee, J. F. Jennings, and E. C. Martin. "Separation of Liquids by Permeation through a Membrane." *Ind. Eng. Chem.* 53 (1961), p. 45.

are furnished. For the material balances of prime importance, as shown in Chapter 4, the data are as follows.

Feed-Reject

Composition: 50 vol % n-heptane; 50 vol % isooctane.
Mass fractions:

$$n\text{-heptane: } \frac{50(0.6883)}{50(0.6883)+50(0.6962)} = \frac{34.42}{34.42+34.81} = 0.497$$

$$\text{isooctane: } \frac{50(0.6962)}{50(0.6883)+50(0.6962)} = 0.503$$

Mole fractions:

$$x_i = \frac{\dfrac{50(0.6883)}{100.2}}{\dfrac{50(0.6883)}{100.2} + \dfrac{50(0.6962)}{114.2}} = \frac{0.3434}{0.3434+0.3048} = 0.530$$

$$x_j = \frac{\dfrac{50(0.6962)}{100.2}}{\dfrac{50(0.6883)}{100.2} + \dfrac{50(0.6962)}{114.2}} = 0.470$$

Specific gravity:

$$\frac{50(0.6883)+50(0.6992)}{100} = 0.6923$$

Molecular weight:

$$100.2(0.530)+114.2(0.470) = 106.8$$

Permeate

Composition: 75 vol % n-heptane; 25 vol % isooctane.
Mass fractions:

$$n\text{-heptane: } \frac{51.62}{51.62+17.41} = 0.748$$

$$\text{isooctane: } \frac{17.41}{51.62+17.41} = 0.252$$

Mole fractions:

$$y_i = \frac{0.5152}{0.5152 + 0.1524} = 0.772$$

$$y_j = 0.228$$

Specific gravity:

$$\frac{75(0.6883) + 25(0.6962)}{100} = 0.6903$$

Molecular weight:

$$100.2(0.772) + 114.2(0.252) = 103.4$$

Pseudocritical Pressures

Feed-Reject: 396.8(0.530) + 372.5(0.470) = 385.4.
Permeate: 396.8(0.772) + 372.5(0.228) = 391.3.
Mean or average value: $(\overline{P}_{pc}) = 388.4$ psia.

Flux Relationship

Total permeate liquid volumetric flux: 0.14 gal/hr-ft^2.
Total permeate molar flux:

$$G = 0.14 \frac{\text{gal}}{\text{hr-ft}^2} \frac{1}{7.48 \, \text{gal/ft}^3} 62.4(0.6903) \text{lb/ft}^3 \frac{1}{103.4 \, \text{lb/mole}}$$

$$= 0.00780 \frac{\text{lb-moles}}{\text{hr-ft}^2}$$

Point permeability relationship for component i:

$$G_i = Gy_i = P_i \frac{P_L x_i - P_V y_i}{\Delta m}$$

where P_i is a composite value for liquid permeation succeeded by vapor permeation and involves the mean or average compressibility factor. Substituting, where $\Delta m = 1$ mil $= 10^{-3}$ in. $= 0.833(10^{-4})$ ft:

$$0.00780(0.772) = P_i \frac{P_L(0.53) - P_V(0.73)}{0.833(10^{-4})}$$

where it will be assumed, for the purposes here, that that $P_L \gg P_V$.

The two cases are as follows, illustrating the determination of the permeability and the diffusivity or diffusion coefficient:

P_L = 30 psia. Solving for P_i,

$$P_i = \frac{0.00780(0.772)(0.833)(10^{-4})}{30(0.53)} = 0.0003153(10^{-4})\frac{\text{lb-moles of } i}{\text{hr-ft}^2\text{-psia/ft}}$$

For the feed-reject, neglecting the permeate,

$$P_r = \frac{30}{385.4} = 0.0778$$

where

$$z = a\ \text{Pr} = 0.17(0.0778) = 0.0132 \qquad \text{and} \qquad \bar{z} \sim \frac{0.0132}{2} = 0.0066$$

Utilizing the mean compressibility factor, the diffusivity becomes

$$D_i = P_i\bar{z}RT = 0.0003153(10^{-4})(0.0066)(10.73)(100+273)(1.8)$$
$$= 0.015(10^{-4})\frac{\text{ft}^2}{\text{hr}}$$

where R = 10.73 in the units of psia-ft³/°R. In the cgs system,

$$D_i = 0.015(10^{-4})\frac{[12(2.54)]^2}{3600} = 0.0039(10^{-4}) = 0.39(10^{-6})\frac{\text{cm}^2}{\text{sec}}$$

P_L = 130 psia. Solving for P_i,

$$P_i = \frac{0.00778(0.772)(0.833)(10^{-4})}{130(0.53)} = 0.0000730(10^{-4})\frac{\text{lb-moles of } i}{\text{hr-ft}^2\text{-psia/ft}}$$

For the feed-reject, neglecting the permeate,

$$P_r = \frac{130}{385.4} = 0.337$$

where

$$z = a\ \text{Pr} = 0.17(0.337) = 0.0573 \qquad \text{and} \qquad \bar{z} \sim \frac{0.0573}{2} = 0.0286$$

Utilizing the mean compressibility factor,

$$D_i = P_i \bar{z} RT = 0.0000730(10^{-4})(0.0286)(10.73)(100 + 273)(1.8)$$

$$= 0.015(10^{-4}) \frac{ft^2}{hr}$$

where again $R = 10.73$ in the units of psia-ft^3/°R. In the cgs system,

$$D_i = 0.015(10^{-4}) \frac{[12(2.54)]^2}{3600} = 0.0039(10^{-4}) = 0.39(10^{-6}) \frac{cm^2}{sec}$$

Comparison

Interestingly, the same result for the diffusivity D_i is obtained for inlet feed-reject pressures of 15 and 115 psig.

If a true steady-state condition exists, then according to the component flux balances,

$$Gy_i = \frac{P_i}{\Delta m}(\bar{P}_{pc})(x_i - y_i) \quad \text{or} \quad y_i = \frac{(P_i/\Delta m)(\bar{P}_{pc})}{G + (P_i/\Delta m)(\bar{P}_{pc})} x_i = K_i x_i$$

$$Gy_j = \frac{P_j}{\Delta m}(\bar{P}_{pc})(x_j - y_j) \quad \text{or} \quad y_j = \frac{(P_j/\Delta m)(\bar{P}_{pc})}{G + (P_j/\Delta m)(\bar{P}_{pc})} x_j = K_j x_j$$

These expressions are of the same form as those derived in Chapter 3, save $G \equiv V''$ and the mean pseudo reduced pressure $(\bar{P}_{pc})$, in the numerator, is the feed-reject pressure P_L and, in the denominator, is the permeate pressure P_V.

Since the experimental data show that $K_i = 75/50 = 1.5$, there is obviously a contradiction with the previously derived relation, which requires that $K_i < 1$. The inference is that the feed-reject pressure should be used in the numerator and the permeate pressure should be used in the denominator, where

$$G = G_i + G_j = Gy_i + Gy_j = (P_i/\Delta m)(P_L x_i - P_V y_i) + (P_j/\Delta m)(P_L x_j - P_V y_j)$$

$$= P_L[x_i(P_i/\Delta m) + x_j(P_j/\Delta m)] - P_V[y_i(P_i/\Delta m) + y_j(P_j/\Delta m)]$$

which is consistent with $P_L > P_V$. Accordingly, the liquid compressibility relationships should be retained.

It is significant to note that Kesting comments about disputes regarding liquid permeation.[9(pp. 278–279)] Some workers have claimed that the principles of gas permeation do not account for the high liquid permeation rates observed. Others counter, "contending that difficulties attendant upon the measurement of pressures and temperatures of saturated vapor, of removing vapor from the product side of the cell, and of maintaining steady-state conditions account for the anomalous results sometimes reported." Nevertheless, Kesting agrees that there is great potential for the separation of organic liquids.

Pure Components

For the permeation of a pure liquid component, $x_i = y_i = 1$, negating the use of the permeation relationship developed for mixtures. Moreover, the averaged pseudo pressure becomes but the critical pressure for the pure component. That is, for a pure component, the pseudocritical pressure is $P_{pc} = P_c$, the critical pressure of the pure component.

However, the permeability relationship may instead be expressed in terms of a mean or averaged compressibility factor $\bar{z}$ for the reject and permeate pressures, as also previously derived:

$$G_i = -D_i \frac{dc_i}{dx} = -D_i \frac{\Delta c_i}{\Delta m} = D_i \frac{1}{\bar{z} RT} \frac{P_L - P_V}{\Delta m} = P_i \frac{P_L - P_V}{\Delta m}$$

This is the flow permeability form, corresponding to the Darcy relationship for flow through porous media. (Alternately, the flow permeability relationship may be developed in terms of the liquid compressibility, as also previously derived.) Here, of course, $P_L > P_V$.

REFERENCES

1. Hoffman, E. J. *Azeotropic and Extractive Distillation.* New York: Interscience, 1964; Huntington, NY: Krieger, 1977.
2. Brown, G. G., A. S. Foust, D. L. Katz, R. Schneidewind, R. R. White, W. P. Wood, G. M. Brown, L. E. Brownell, J. J. Martin, G. B. Williams, J. T. Banchero, and J. L. York. *Unit Operations.* New York: Wiley, 1950.
3. Katz, D. L., D. Cornell, R. Kobayashi, F. H. Poettmann, J. A. Vary, J. R. Elenbaas, and C. F. Weinaug. *Handbook of Natural Gas Engineering.* New York: McGraw-Hill, 1959.
4. Hoffman, E. J. *Phase and Flow Behavior in Petroleum Production.* Laramie, WY: Energon, 1981.
5. Hwang, S.-T., and K. Kammermeyer. *Membranes in Separations,* vol. VII of *Techniques of Chemistry,* ed. A. Weissberger, Chapters 4 and 8, Appendix B. New York: Wiley-Interscience, 1975.
6. Rautenbach, R. "Process Design and Optimization." In *Handbook of Industrial Membrane Technology,* ed. M. C. Potter. Park Ridge, NJ: Noyes Publications, 1990.

7. Hoffman, E. J. *Unsteady-State Fluid Flow: Analysis and Applications in Petroleum Reservoir Behavior.* Amsterdam, the Netherlands: Elsevier Science, 1999.

8. *International Critical Tables.* New York: McGraw-Hill, 1926–1930.

9. Kesting, Robert E. *Synthetic Polymeric Membranes*, p. 281. New York: McGraw-Hill, 1971.

10. Hougen, O. A., K. M. Watson, and R. A. Ragatz. *Chemical Process Principles*, part II, *Thermodynamics.* New York: John Wiley & Sons, 1959.

11. Hoffman, E. J. *Analytic Thermodynamics: Origins, Methods, Limits, and Validity.* New York: Taylor and Francis, 1991.

12. Lee, S. Y., and B. S. Minhas. "Effect of Gas Composition on Permeation through Cellulose Acetate Membranes." In *New Membrane Materials and Processes for Separation*, eds. Kamalesh K. Sirkar and Douglas R. Lloyd. AIChE Symposium Series vol. 84, no. 261, 1987 AIChE Summer National Meeting, Minneapolis. New York: American Institute of Chemical Engineers, 1988.

3

Single-Stage Membrane Separations

Consider the following membrane stream juxtaposition, by analogy with a phase separation, as diagrammed in Figure 3.1. The conditions and compositions for each stream do not change with position; the circumstance is called *perfect mixing*, and the conditions and compositions do not change with time, signifying a steady state.

The mole fraction compositions y_i and x_i are therefore uniform on each side of the membrane, where the subscript i denotes components 1, 2, 3, ..., k. The respective steady-state molar stream rates are denoted F, L, and V. These may designate the total flow rate of the each stream or be a flux rate based on the membrane area.

Stream F denotes the feed, stream V the permeate, and stream L the reject. Ordinarily, all phases are to be gaseous, but alternately all may be liquids; that is, no phase separations per se are involved. Furthermore, the system is nonreacting. The kinds of calculations involved are presented in a number of references, as applicable to phase separations, at equilibrium, between liquids and gases or vapors.[1,2,3] By a fortuitous circumstance in the representations, the same methodology can be applied to membrane separations.

The material balances are

$$F = L + V$$
$$F(x_F)_i = Lx_i + Vy_i$$

where

$$\sum (x_F)_i = 1 \qquad \sum x_i = 1 \qquad \sum y_i = 1$$

Furthermore,

$$(L + V)(x_F)_i = Lx_i + Vy_i$$

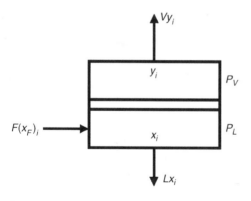

Figure 3.1 Single-stage membrane separation with perfect mixing.

whereby

$$\frac{L}{V} = \frac{y_i - (x_F)_i}{(x_F)_i - x_i}$$

or

$$y_i = -\frac{L}{V}x_i + \left[\frac{L}{V} + 1\right](x_F)_i$$

This is a straight line in y_i–x_i space, with the slope $(-L/V)$ and the y intercept at $[L/V + 1](x_F)_i$ for constant parameters of V, but which in general is a variable.

The membrane rate balance is, for each component i,

$$Vy_i = P_i(P_L x_i - P_V y_i)A$$

where here

P_i = overall membrane permeability to component i in moles per unit time per unit area per unit pressure difference or partial pressure difference;
P_L = system pressure on the high side;
P_V = system pressure on the low side;
A = membrane area perpendicular to flow; preferably based on the permeate side.

It is understood that the feed pressure P_F is higher than or approximately equal to P_L, so that the flow is sustained.

The foregoing rate equation can alternately be expressed in terms of the permeate flux, designated as $V'' = V/A$ or as $G = V/A$. Therefore, in consistent units,

$$V''y_i = P_i(P_L x_i - P_V y_i)$$

As a further note, the membrane permeability to a component i is most likely determined experimentally, using only pure component i. In the application to mixtures, the projection is made that the same value for the permeability can be used if the driving force is in terms of the partial pressure of the component i.

It may be further observed that

$$\sum V''y_i = V'' \qquad \sum y_i = V'' = P_L \qquad \sum P_i x_i - P_V \qquad \sum P_i y_i$$

This serves as an expression for the variable $V'' = V/A$ in terms of y_i and x_i.

Since $F = V + L$ and $F(x_F)_i = Vy_i + Lx_i$ it follows that either V or y_i or x_i can be eliminated as a variable or variables.

Note also that

$$y_i = Vy_i/V = V''y_i/V'' = (P_L P_i x_i - P_V P_i y_i) \Big/ \left(P_L \sum P_i x_i - P_V \sum P_i y_i \right)$$

Hence, the mole fraction can also be expressed in terms of pressure and permeability.

3.1 TERMS AND UNITS

It is understood that, if the membrane permeability is in the units of moles/time, then the flow rate L is in moles per unit time per unit membrane area. (The other flow rates would also be based on membrane area.) In turn, the notation L may be replaced by the flux G_L; that is, the flow rate may be placed on an areal basis.

If the units of permeability are in volume per unit time, then the flow rates are in volume per unit time, though adjustments or accommodations must be made for the pressure (and temperature).

If the total permeate flow rate is given by V, say, the component flow rate can be designated as

$$V_i = Vy_i$$

whereas the corresponding flux rate for component i in the permeate phase can be written as $(G_V)_i = (G_V)y_i = V''y_i$. For the flux of component i in any phase, in general, the symbol G_i would suffice.

Membrane permeability is customarily based on pressure drop per unit membrane thickness. The overall permeability then becomes the permeability as per unit thickness divided by the thickness. Therefore, as the membrane thickness increases, the overall permeability decreases.

As previously stated, the units commonly used for pointwise membrane permeability (or mobility) are

$$10^{-9} \text{ cm}^3/\text{sec-cm}^2\text{-cm Hg/cm}$$

where the unit term cm Hg/cm represents the partial pressure drop in centimeters of mercury per unit membrane thickness. The volume in cm^3 is at standard conditions.

The relative permeability, one component to another, has also been defined as the selectivity α. Therefore,

$$\alpha = \alpha_{i-j} = P_i/P_j$$

is the permeability of component i relative to component j. The term α will not be further employed as such.

A relative permeation flux φ or φ_{i-j} may be defined for the permeate phase as

$$\varphi \text{ or } \varphi_{i-j} = (G_V)_i/(G_V)_j = V''y_i/V''y_j = Vy_i/Vy_j$$

The relative permeation flux, in general, differs from the selectivity and depends on the composition as well as the pressure. It is also a term that will not be employed further.

3.2 MOLE FRACTION RELATIONSHIPS

It follows from rearranging the rate balance for a component i that

$$y_i = \frac{P_i P_L}{V'' + P_i P_V} x_i$$

or

$$y_i = K_i x_i$$

where K_i is defined by the substitution. This represents a straight line in y_i–x_i space with slope K_i, extending from the origin for constant parameters of the variable V''. It is of the same notation and symbolism as the equilibrium vaporization ratio or K-value encountered in the representation of vapor-liquid equilibria and may be called the *permeation coefficient* or *distribution coefficient* for component i.

It may be observed that the units for V'' are the same as for both V/A and $P_i P_V$, where in its usage P_i refers to the overall permeability rather than the pointwise permeability; that is, strictly speaking, the comparison of units is as follows, for V'' versus $P_i P_V$:

$$\frac{\text{total moles}}{\text{time-area}} \text{ vs. } \frac{\text{moles of } i}{\text{time-area-}pp_i}(\text{total pressure})$$

$$= \frac{\text{moles of } i}{\text{time-area-(total pressure)}\dfrac{\text{moles of } i}{\text{total moles}}}(\text{total pressure}) = \frac{\text{total moles}}{\text{time-area}}$$

where for convenience pp_i denotes the partial pressure of component i.

Note in particular that the ratios of values as just represented is not the selectivity, as would be defined by the permeability ratios; that is, by $\alpha_{i-j} = P_i/P_j$. Furthermore, the selectivity as used here is different than the concept of relative volatility, which would be the ratio of the K-values, one to another. Furthermore, the ratio of the K-values so determined is lower than would be suspected from the ratio of the permeabilities. The implication is that the ensuing permeability separations are much less sharp than would be suspected from the permeability ratios or selectivities.

In other words, the presence of the parameter or variable V'', notably, affects these ratios of K-values; that is,

$$\frac{K_i}{K_j} = \frac{P_i P_L}{P_j P_L}\frac{V'' + P_i P_V}{V'' + P_j P_V} = \frac{P_i}{P_j}\frac{1 + P_j(P_V/V'')}{1 + P_i(P_V/V'')}.$$

Furthermore, the relatively larger is the value of V'', the more likely that the K ratio, as designated previously, approximates the permeability ratio. The relatively smaller is the value of V'', the more likely that the K ratio approaches unity. Last, the K ratio must be greater than unity for a separation to occur.

The foregoing provides a prime reason for the fact that single-phase or pure component permeabilities do not necessarily pertain to mixtures, as noted and referenced at the end of Section 2.2 of Chapter 2. That is, the effective permeability for the components in mixtures tend to be less, or much less, than for the permeability of the pure components determined alone.

The parameter or variable V'' has the dimensions of permeability times pressure, as previously observed in terms of moles. However, in terms of the total gas volume permeated, using the pointwise permeability and dealing with volume fractions, V'' has the dimensions of, say,

$$\frac{\text{cm}^3}{\text{cm}^2\text{-sec-}\dfrac{1}{\text{cm}}} = \frac{\text{cm}^2}{\text{sec}}$$

which, interestingly, are the customary units for the diffusion coefficient or diffusivity.

It should be further emphasized, however, that, if the permeability is expressed as the overall permeability (by dividing by the membrane thickness in cm), then V'' has the net dimensions of velocity:

$$\frac{cm^3}{cm^2\text{-sec}} = \frac{cm}{sec}$$

If the volumetric permeate flow in standard cm^3 is converted to g-moles (by dividing by 22,414 standard cm^3/g-mole), the dimensions, of course, are

$$\frac{\text{g-moles}}{cm^2\text{-sec}}$$

which are the dimensions for molar flux. In this way, say, the membrane area can be related to, or determined from, the molar flux; that is, $A = V/V''$ in consistent units, as is illustrated in an example at the end of the chapter.

Bubble-Point Type Determination

Note that, when $V/F \rightarrow 0$, it is required that

$$\sum y_i = 1 = \sum K_i x_i = \sum \frac{P_i P_L}{V'' + P_i P_V} x_i$$

This would correspond to the bubble-point calculation as performed for vapor-liquid equilibrium, the object being to determine the temperature at a given pressure, or vice versa, whereby the first "drop" of vapor ensues from the vaporization of the liquid phase; that is, it would correspond to a point or locus of points on the saturated liquid curve.

Here, however, the situation corresponds to the circumstance where the first "drop" of permeate ensues. Or, if the permeate rate is to be finite, then both the feed and reject must be infinite or increase without limit. In other words, all the feedstream is rejected, albeit an infinitely small amount of permeate phase would be produced.

Observe that the composition of the permeate is y_i, whereas the composition of the reject x_i is the same as that of the feed.

Dew-Point Type Determination

Alternately,

$$\sum x_i = 1 = \sum \frac{y_i}{K_i} = \sum \frac{V'' + P_i P_V}{P_i P_L} y_i$$

This corresponds to the dew-point calculation as performed for a vapor-liquid equilibrium condition; that is, it corresponds to a point or locus of points on the saturated vapor curve as distinguished from the saturated liquid curve. (For a single or pure component, they are the same.)

In permeation, however, this corresponds to the case where all the feedstream goes through the membrane; hence, the permeate rate equals the feed rate and the reject rate is nil—albeit the composition x_i pertains to the (infinitesimal) "drop" of reject produced whereas the composition y_i is the same as that of the feed.

It may be added, however, that these representations and calculations pertain to nonequilibrium behavior for the membrane permeation of the components of gaseous systems. The same sort of notation may be adapted to liquid systems.

Transient vs. Steady-State Behavior in Permeability Determinations

The foregoing derivations raise some intriguing speculations about the measurement and determination of permeability for the respective components in a mixture. If a true or complete steady-state condition exists during the experiment, where all of the feedstream passes through the membrane, then the ratios $V/F = 1$ and $L/F = 0$; that is, it can be said that no reject phase is produced.

Furthermore, when $V/F = 1$, no finite separation occurs, albeit a dew-point type calculation gives a value for the degree or sharpness of separation in terms of mole fraction ratios or K-values. When $V/F \to 0$, again no finite separation occurs, albeit a bubble-point type calculation gives a value for the degree or sharpness of separation in terms of mole fraction ratios or K-values. (It may be added that, for a single pure component, whether or not a reject phase can be said to exist is of no concern, since V/F and L/F do not enter into the determination and calculations.)

However, in actual test measurements, at what point, if any, can it be said that all the feed passes through the membrane? That is to say, does not holdup occur on the upstream pressure side? For in any kind of short-term or transient test (say, in what might be called a batch or semi-continuous laboratory or bench-scale test), does a reject phase not exist at any point? At any point in time, is there no situation in which the feed that has not yet passed through the membrane constitute a reject phase? Only for a long-term, steady-state test—with no reject sidestream—can it truly be said that all the feed passes through the membrane. This sort of long-term test, properly speaking, then provides the true measure of membrane permeability for the components within a mixture. Whether or not discrepancies therefore exist between the results of short-term tests and long-term tests is an interesting philosophical question.

In any event, the permeability determinations of components in mixtures are apparently at variance with those determined separately for each of the pure components. This general problem is often encountered in trying to project from pure component behavior to the behavior of mixtures.

Unit Permeation Rate

The expression for the K-value can conveniently be rewritten as

$$K_i = \frac{P_i P_L}{V'' + P_i P_V} = \frac{P_i P_L / P_i P_V}{V^* + 1} = \frac{P_L / P_V}{V^* + 1}$$

where $V^* = V'' / P_i P_V$ can be called the *dimensionless* or *reduced permeation flux*, or some other designator can be used.

It may again be observed that, since K is dimensionless, the units of the molar flux V'' are to be in the same units as the feed molar flux F'' and in the same units as the combination $P_i P_V$ or $P_i P_L$. Similarly, V is in the same units as the molar feed rate F. These combined units may be in, say, cc per unit time (at standards conditions) or moles per unit time, and so forth; that is, the areal basis can pertain to the entire membrane or membrane assembly. Accordingly, the permeability P_i can pertain to the entire membrane per se.

Alternately, P_i can, of course, be placed on a unit area basis (e.g., per square centimeter). In turn, the feed rate F, permeate rate V, and reject rate L then are on the same common unit area basis. For the further purposes here, the K-value calculations utilize V'' rather than V^*, inasmuch as V'' more directly stands for the permeate flux in multistage operations.

Expected vs. Actual Separations

As previously indicated, the permeability values within mixtures are generally less than those for the pure components. Or, the degree of separation in mixtures is less sharp than expected from the permeability of the pure components. This can perhaps be traced to projecting the idea of relative volatility to membrane relative permeability or selectivity.

The concept of relative volatility in vapor-liquid equilibria can be expressed as

$$\alpha_{i-r} = \frac{K_i}{K_r} \qquad \alpha_{j-r} = \frac{K_j}{K_r}$$

where r denotes some reference component. Therefore,

$$y_i = K_i x_i = K_r \alpha_{i-r} x_i \qquad y_j = K_j x_j = K_r \alpha_{j-r} x_j$$

Since $y_i + y_j = 1$, then

$$K_r = \frac{1}{\alpha_{i-r} x_i + \alpha_{j-r} x_j} = \frac{1}{\sum \alpha_{i-r} x_i}$$

In turn,

$$y_i = \frac{\alpha_{i-j} x_i}{\sum \alpha_{i-j} x_i}$$

If membrane permeability or selectivity is introduced in lieu of relative volatility, then the effect would seemingly enhance the separation, as follows:

$$y_i = \frac{(P_i/P_r) x_i}{\sum (P_i/P_r) x_i} = \frac{P_i x_i}{\sum_i P_i x_i} = K_i x_i$$

That is, the equivalent K-value for the more permeable component would seem higher and the equivalent K-value for the less permeable component would seem lower. This effect is indicated, for instance, by a comparison made in Example 3.1 but is, of course, not rigorous.

3.3 MULTICOMPONENT SEPARATION CALCULATIONS

In general, for any circumstance, since

$$\begin{aligned} F(x_F)_i &= V y_i + L x_i \\ &= V K_i x_i + L x_i \\ &= V y_i + L y_i/K_i \end{aligned}$$

then

$$\sum \frac{(x_F)_i}{\dfrac{V}{F} K_i + \dfrac{L}{F}} = \sum x_i = 1$$

or

$$\sum \frac{(x_F)_i}{\dfrac{V}{F} + \dfrac{L}{F}\dfrac{1}{K_i}} = \sum y_i = 1$$

where

$$\frac{V}{F} + \frac{L}{F} = 1$$

$$K_i = \frac{P_i P_L}{V'' + P_i P_V} = \frac{P_i P_L / P_i P_V}{V^* + 1} = \frac{P_L / P_V}{V^* + 1}$$

Given the $(x_F)_i$, then each assumed value V/F (or L/F) has a unique solution for V''. This is a variation on the single-stage flash calculation for a vapor-liquid separation.

The calculation for a multicomponent system is, in general, trial and error, establishing the values of x_i and the y_i along with a corresponding value for V''. In turn, given the feed rate F and the specified ratio V/F, the absolute value of the permeate rate V with respect to F follows; similarly for determining an absolute value for the reject rate L relative to F.

As the limiting conditions, note that, if $V/F = 0$ and $L/F = 1$, then

$$\sum K_i (x_F)_i = \sum y_i = 1$$

and if $V/F = 0$ and $L/F = 1$, then

$$\sum (x_F)_i / K_i = \sum x_i = 1$$

Given the value of $(x_F)_i$, these calculations establish the respective values of V'' for each of the limiting conditions, along with the respective compositions x_i and y_i. These limiting bubble-point and dew-point type determinations were previously described.

Key Components

In the parlance used for distillation calculations, the two key components can be designated as those whose distribution behavior is closest to unity, with one key component showing a K-value less than 1 and the other greater than 1. The latter would exhibit the greater "volatility" or activity, in this case, would have a greater value for K.

Therefore, if

$$K_i = \frac{P_i P_L}{V'' + P_i P_V} > 1$$

$$K_j = \frac{P_j P_L}{V'' + P_j P_V} < 1$$

then *i* would be perceived as the more "volatile" or active component and *j* the lesser.

Note further that, if $K_i > K_j$, then

$$\frac{P_i P_L}{V'' + P_i P_V} - \frac{P_j P_L}{V'' + P_j P_V} > 0$$

or

$$V'' P_i P_L + P_i P_j P_L P_V - V'' P_j P_L - P_i P_j P_L P_V > 0$$

Collecting terms,

$$V'' P_L (P_i - P_j) > 0 \quad \text{or} \quad (P_i - P_j) > 0$$

That is, if $P_i > P_j$, then component *i* would have the greater permeability and have the higher "volatility" or activity.

Extra Constraints

As a special case, let $K_i K_j = 1$. It would then follow that

$$\frac{P_i P_j P_L^{\ 2}}{(V'' + P_i P_V)(V'' + P_j P_L)} = 1$$

or

$$0 = \left(V''\right)^2 + \left(P_i P_V + P_j P_V\right)V + \left[-P_i P_j \left(P_L^{\ 2} - P_V^{\ 2}\right) \right]$$

or

$$0 = a(V'')^2 + bV'' + c$$

and

$$V'' = \frac{-b \pm \sqrt{b^2 - 4ac}}{2a}$$

where the quantities are defined by the substitutions. This, in general, is not true, however, even for a two-component system.

In fact, the foregoing introduces a contradiction, since the mole fraction summation cannot be satisfied at the same time. Therefore, an a priori constraint cannot be introduced between K_i and K_j; that is, no additional equation can be introduced. And, if so, it would pertain only to a particular situation; that is, some unique combination of the variables.

Consider the dew-point type of calculation. If, for components 1 and 2,

$$\frac{(x_F)_1}{K_1} + \frac{1-(x_F)_1}{K_2} = 1$$

then, if $K_2 = 1/K_1$,

$$K_2(x_F)_1 + K_1 - K_1(x_F)_1 = K_2 K_1$$

$$(x_F)_1 + K_1^2 - K_1^2(x_F)_1 = K_1$$

or

$$K_1^2[1-(x_F)_1] = K_1 - (x_F)_1$$

Therefore, solving the quadratic for K_1,

$$K_1 = \frac{1 \pm \sqrt{1 - 4[1-(x_F)_1](x_F)_1}}{2[1-(x_F)_1]}$$

In other words, K_1 is required to take on this unique value if $K_2 = 1/K_1$.

3.4 TWO-COMPONENT CALCULATIONS

A simplification can be made for binary systems. For two components i and j, let

$$\frac{(x_F)_i}{V/F + (1-V/F)\left[\dfrac{V''+a}{b}\right]} = \frac{(x_F)_j}{V/F + (1-V/F)\left[\dfrac{V''+c}{d}\right]}$$

where

$$a = P_i P_V$$
$$b = P_i P_L$$
$$c = P_j P_V$$
$$d = P_j P_L$$

This may be further arranged as

$$\frac{\dfrac{b(x_F)_i}{(1-V/F)}}{\left[\dfrac{V/F}{1-V/F}b+a\right]+V''} + \frac{\dfrac{d(x_F)_j}{(1-V/F)}}{\left[\dfrac{V/F}{1-V/F}d+c\right]+V''} = 1$$

or

$$\frac{(\overline{x}_F)_i}{\alpha + V''} + \frac{(\overline{x}_F)_j}{\beta + V''} = 1$$

where the introduced quantities are defined by the substitutions. Accordingly,

$$\beta(\overline{x}_F)_i + V''(\overline{x}_F)_i + \alpha(\overline{x}_F)_j + V''(\overline{x}_F)_j = \alpha\beta + V''(\alpha + \beta) + (V'')^2$$

or

$$0 = (V'')^2 + \{(\alpha + \beta) - [(\overline{x}_F)_i + (\overline{x}_F)_j]\}V'' + (-[\beta(\overline{x}_F)_i + \alpha(\overline{x}_F)_j] + \alpha\beta\}$$

and which can be represented as

$$0 = A(V'')^2 + BV'' + C$$

Therefore, solving the quadratic for V'',

$$V'' = \frac{-B \pm \sqrt{B^2 - 4AC}}{2A}$$

where the quantities are defined by the substitutions, with $A = 1$, and where

$$\alpha = \frac{V/F}{1 - V/F} b + a$$

$$\beta = \frac{V/F}{1 - V/F} d + c$$

$$(\overline{x}_F)_i = \frac{b}{1 - V/F}(x_F)_i$$

$$(\overline{x}_F)_j = \frac{d}{1 - V/F}(x_F)_j$$

The quantity B in the quadratic is positive and the $\pm$ sign is used as its plus value.

The calculation is readily performed for the condition $V/F \to 0$, analogous to the bubble-point type determination. If, however, $V/F \to 1$, then the dew-point type determination must be used, so that

$$1 = \sum x_i = \sum y_i / K_i$$

or

$$1 = \frac{V'' + a}{b}(\overline{x}_F)_i + \frac{V'' + c}{d}(\overline{x}_F)_j$$

where

$$bd = V''d(x_F)_i + ad(x_F)_i + V''b(x_F)_i + bc(x_F)_i$$

where the constants a, b, c, and d have been previously identified. Collecting terms and solving for V'',

$$V'' = \frac{bd - [ad(\overline{x}_F)_i + bc(\overline{x}_F)_i]}{d(\overline{x}_F)_i + b(\overline{x}_F)_i}$$

where, as noted, the quantities have previously been defined.

3.5 EFFECT OF RECYCLE

If a recycle stream R is introduced, as shown in Figure 3.2, then for a component i,

$$F(x_F)_i + Ry_i = (V + R)y_i + Lx_i$$
$$(V + R)y_i = P_i(P_Lx_i - P_Vy_i)A$$

or

$$(V'' + R'')y_i = P_i(P_Lx_i - P_Vy_i)$$

where $V'' = V/A$ and $R'' = R/A$.

The overall material balance remains the same since R (or R'') cancels out:

$$F(x_F)_i = Vy_i + Lx_i$$

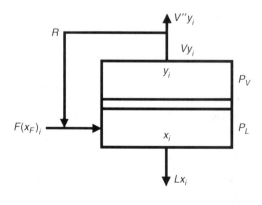

Figure 3.2 Recycle configuration.

However, the membrane rate balance appears as

$$y_i = \frac{P_i P_L}{(V'' + R'') + P_i P_V} x_i$$

$$= \frac{P_i P_L}{V' + P_i P_V} x_i = K'_i x_i$$

Therefore,

$$\sum \frac{(x_F)_i}{V/F + (1 - V/F)(1/K'_i)} = \sum y_i = 1$$

where, as noted,

$$K'_i = \frac{P_i P_L}{(V'' + R'') + P_i P_V} = \frac{P_i P_L}{V' + P_i P_V}$$

The solution procedure is the same except that V'' is replaced by $V' = V'' + R''$ and K_i by $K'_i K'_i$, where $V'' = V' - R''$.

The effect is to decrease the net membrane capacity. The separation remains the same; that is, as pertains to the foregoing interpreted conditions for perfect mixing.

Similarly, in equilibrium flash calculations, the recycle of the liquid or vapor phase has no effect, ideally, on the phase compositions. The only effect is on the net capacity of the phase separator.

EXAMPLE 3.1

Various and random membrane information has been tabulated as a matter of course in Chapters 1 and 2. For the calculation purposes here, a representative set of comparative values follows for a membrane of low selectivity between components i and j, with operating pressure levels in the ratio of 3/2. The units are unstated, inasmuch as the entities calculated absorb the conversion factors, which are not necessary for calculating the degree of separation and therefore immaterial save in determining membrane area.

Membrane Data and Operating Data

$$P_i = 20 \qquad a = P_i P_V = 40$$
$$P_j = 10 \qquad b = P_i P_L = 60$$
$$P_L = 3c = P_i P_V = 20$$
$$P_V = 2d = P_j P_L = 30$$
$$K_i = 60/(V'' + 40), \text{ where } 10 < V'' < 20$$
$$K_j = 30/(V'' + 20)$$

Feed Composition

$$(x_F)_i = 0.4$$
$$(x_F)_j = 0.6$$

Substitutions

$$\alpha = \frac{V/F}{1 - V/F}(60) + 40, \quad \text{where } b = 60 \quad a = 40$$

$$\beta = \frac{V/F}{1 - V/F}(30) + 20, \quad \text{where } d = 30 \quad c = 20$$

$$(\overline{x}_F)_i = \frac{60}{1 - V/F}(0.4) = \frac{24}{1 - V/F}$$

$$(\overline{x}_F)_j = \frac{30}{1 - V/F}(0.6) = \frac{18}{1 - V/F}$$

Furthermore,

$$x_i = \frac{(\overline{x}_F)_i}{\alpha + V''} \qquad x_j = \frac{(\overline{x}_F)_j}{\alpha + V''}$$

$$K_i = \frac{b}{V'' + a} \qquad K_j = \frac{d}{V'' + c}$$

$$y_i = K_i x_i \qquad y_j = K_j x_j$$

Dew-Point Type Determination ($V/F = 1$)

$$V'' = \frac{60(30) - [40(30)(0.4) + 60(20)(0.6)]}{30(0.4) + 60(0.6)}$$

$$= \frac{1800 - [480 + 720]}{12 + 36} = \frac{1800 - 1200}{48} = 12.5$$

$$K_i = \frac{60}{12.5 + 40} = 1.1429 \qquad K_j = \frac{30}{12.5 + 20} = 0.9231$$

$$x_i = 0.4/1.1429 = 0.350$$

$$x_j = 0.6/0.9231 = 0.650$$

for a total of x_i and x_j of 1.000.

Calculation Sequence

The calculational sequence is provided in Tables 3.1–3.4 for the range of values of V/F.

Note: The V/F ratio is a process variable or parameter to be affixed by the operator. Furthermore, it can be assumed that P_L and P_V are set by back-pressure controllers on gas streams L and V. The feed rate F may be increased or reduced by a valve in the line, such as by a flow controller, where the upstream feed pressure is sufficiently high. Ordinarily, it would be set at a constant rate, at a fixed reject pressure P_L.

In turn, the rates L and V adjust to the pressure difference maintained across the membrane, which is also related to the membrane permeability. If a higher permeate rate is desired, then the pressure P_V must be lowered and or the feed rate F increased. (Alternately, albeit it is not a process control variable, the membrane surface or size can be increased.) It should be emphasized, moreover, that the calculations are process design estimations, prior to fabrication and operation. The corresponding instrumentation schematic is shown in Figure 3.3.

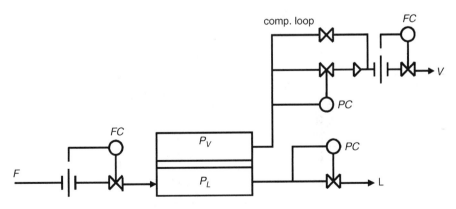

Figure 3.3 Instrumentation schematic.

Table 3.1 Calculation of Constants

V/F	α	β	$(\bar{x}_F)_i$	$(\bar{x}_F)_j$	$\alpha\beta$	$\alpha + \beta$	$(\bar{x}_F)_i + (\bar{x}_F)_j$	B	$\alpha(\bar{x}_F)_i$	$\beta(\bar{x}_F)_j$	C
0.0	40	20	24	18	800	60	42	18	480	720	−400.00
0.1	46.67	23.333	26.67	20	1,088.81	70	46.67	23.33	622.22	933.34	−466.75
0.2	55	27.50	30	22.5	1,512.50	82.5	52.5	30	825	1,237.5	−550
0.3	65.714	32.857	34.286	25.714	2,159.16	98.571	60	38.571	1,126.54	1,689.77	−657.15
0.4	80	40	40	30	3,200	120	70	50	1,600	2,400	−800
0.5	100	50	48	36	5,000	150	84	66	2,400	3,600	−1,000
0.6	130	65	60	45	8,450	195	105	90	3,900	5,850	−1,300
0.7	180	90	80	60	16,200	270	140	130	7,200	10,800	−1,800
0.8	280	140	120	90	39,200	420	210	210	16,800	25,200	−2,800
0.9	580	290	240	180	168,200	870	420	450	69,600	104,400	−5,800
1.0	—	—	—	—	—	—	—	—	—	—	—

Table 3.2 Calculation of V'' and K_i

V/F	V''	K_i	K_j	K_iK_j
0.0	12.9317	1.133536	0.910977	1.0326
0.1	12.8874	1.134485	0.912203	1.0349
0.2	12.8388	1.135529	0.913553	1.0373
0.3	12.7938	1.136500	0.914807	1.0397
0.4	12.7493	1.137458	0.916053	1.0420
0.5	12.7056	1.138399	0.917274	1.0443
0.6	12.6628	1.139325	0.918476	1.0465
0.7	12.6209	1.140231	0.919656	1.0487
0.8	12.5798	1.141123	0.920816	1.0508
0.9	12.5395	1.141998	0.92196	1.0529
1.0	12.5000	1.142857	0.92308	1.0530

Table 3.3 Calculation of Phase Compositions (of Permeate V and Reject L)

V/F	$\alpha + L$	y_i	$1/K_i$	x_i	$\beta + L$	y_j	$1/K_j$	x_j
0.0	52.9317	0.4534	0.8829	0.4000	32.9317	0.5466	1.0977	0.6000
0.1	59.5574	0.4478	0.8815	0.3947	36.2207	0.5522	1.0962	0.6053
0.2	67.8388	0.4422	0.8806	0.3894	40.3388	0.5578	1.0946	0.6106
0.3	78.5078	0.4367	0.8799	0.3843	45.6508	0.5633	1.0931	0.6157
0.4	92.7492	0.4313	0.8792	0.3792	52.7492	0.5687	1.0916	0.6208
0.5	112.7056	0.4259	0.8784	0.3741	62.7056	0.5741	1.0902	0.6259
0.6	142,6628	0.4206	0.8777	0.3692	77.6628	0.5794	1.0888	0.6309
0.7	192.6209	0.4153	0.8770	0.3642	102.6209	0.5847	1.0874	0.6358
0.8	292.5798	0.4101	0.8763	0.3594	152.5798	0.5899	1.0860	0.6406
0.9	592.5395	0.4050	0.8757	0.3546	302.5395	0.5950	1.0847	0.6454
1.0	—	0.4000	0.8750	0.3500	—	0.6000	1.0833	0.6500

Table 3.4 Separations and Recoveries

V/F	$y_i/x_i = K_i$	$Vy_i/F(x_F)_i$	L/F	$x_j/y_j = 1/K_j$	$Lx_j/F(x_F)_j$
0.0	1.1335	0.0000	1.0	1.0977	1.0000
0.1	1.1345	0.1120	0.9	1.0962	0.9080
0.2	1.1356	0.2211	0.8	1.0946	0.8141
0.3	1.1364	0.3275	0.7	1.0931	0.7183
0.4	1.1374	0.4313	0.6	1.0916	0.6208
0.5	1.1385	0.5324	0.5	1.0902	0.5216
0.6	1.1392	0.6309	0.4	1.0888	0.4206
0.7	1.1403	0.7268	0.3	1.0874	0.3179
0.8	1.1411	0.8202	0.2	1.0860	0.2135
0.9	1.1421	0.9113	0.1	1.0847	0.1083
1.0	1.1429	1.0000	0.0	1.0833	0.0000

Relative Volatility-Type Calculation
If it is assumed that

$$y_i = \frac{P_i x_i}{\sum P_i x_i}$$

then, based on the feed composition,

$$y_i = \frac{20}{20(0.4) + 10(0.6)} x_i \quad \text{or} \quad y_i = 1.428571 x_i$$

$$y_j = \frac{10}{20(0.4) + 10(0.6)} x_i \quad \text{or} \quad y_j = 0.714286 x_j$$

These values for K diverge considerably from those appearing in Table 3.2 (say for $V/F = 0$), indicating a much sharper separation than obtained via the rigorous calculations involving V.

Calculation of Membrane Area
The separation calculations have not required any units for permeability and pressure or pressure difference nor for membrane thickness. Accordingly, units now are assumed with the following numbers:

$P_i = 20\ (10^{-9})\ \text{cm}^3/\text{cm}^2\text{-sec-cm Hg/cm}$

$P_j = 10$

$P_L = 30$ atm

$P_V = 20$ atm

$\Delta m = 10$ microns or $10(10^{-4})$ cm, the membrane thickness

Accordingly, to convert from the dimensionless properties supplied, the conversion factor for the dimensionless flux value V'' in Table 3.2 is as follows. As previously derived,

$$V'' = \frac{V}{A} = G = P_L \sum P_i x_i - P_V \sum P_i y_i$$

where the summation is for both components i and j. It follows that, for the units specified, the corresponding value of V'' as calculated in Table 3.2 must have the following units, if pressure is in atmospheres:

$$\frac{(10^{-9})\text{cm}^3}{\text{cm}^2\text{-sec-cm Hg/cm}}\ \text{atm}$$

It may be noted that a pointwise permeability is specified, which must be divided by the membrane thickness to obtain the overall permeability. The foregoing units may be converted to more convenient units by multiplying V'' as previously calculated in Table 3.2, as follows:

$$V'' \times \frac{76 \text{ cm Hg}}{\text{atm}} \quad \frac{1}{(\Delta m \text{ in cm})} \quad \frac{(10/1)}{\{(22,414(10^9)) \text{ in } [(10^{-9})\text{cm}^3]/(g-\text{mole})\}}$$

$$= V'' \times \frac{76}{22,414} \frac{10^{-8}}{(\Delta m \text{ in cm})}$$

where the factor $(10/1)$ denotes that the membrane pressures have been converted from a nominal 3 and 2 to 30 and 20 atm, differing by a multiple of 10 in this case.

That is to say, in the term in braces in the denominator, there would be the number $22,414(10^9)$ measured in the units of $[(10^{-9}) \text{ cm}^3]$ of gas per g-mole. (Which of course is identical to $22,414 \text{ cm}^3$ of gas per g-mole.) Significantly, however, the units of $(10^{-9})\text{cm}^3$ cancel out with these same units as occurring in P_i.

The foregoing convoluted conversion of units gives a new value for the permeate flux V'' in the following units:

$$\frac{\text{g-moles}}{\text{cm}^2\text{-sec}}$$

The value so obtained can be placed on the basis of g-mole/sec of feed-stream F. Therefore, the corresponding area requirement for each value of V'' in Table 3.2 is

$$A = \frac{V}{F} \frac{10^9}{V'' \dfrac{76}{22,414} \dfrac{(10/1)}{(\Delta m \text{ in cm})}} = \frac{V}{F} \frac{1}{V''} \frac{1}{0.00339074} \frac{(\Delta m \text{ in cm})(10^9)}{10}$$

For a membrane thickness of 10 microns or $10(10^{-4})$ cm, this transforms to

$$A = \frac{V}{F} \frac{1}{V''} \frac{1}{0.00339074} \frac{(10)(10^{-4})(10^9)}{10} = \frac{V}{F} \frac{1}{V''} \frac{1}{0.00339074(10^{-5})}$$

$$= \frac{V}{F} \frac{1}{V''} \frac{1}{3.39074(10^{-8})}$$

where, in this case, the number $3.39074(10^{-8})$ can be treated as a conversion factor. The area requirement so obtained is in cm^2 per g-mole of feed per second. Note that 929 cm^2 is 1 ft^2.

For a value, say, of $V'' \sim 12.7$, where $V/F = 0.5$, and $\Delta m = 10(10^{-4})$ cm as stipulated, the area calculates to $1.16(10^6)$ cm^2 or 1,250 ft^2 for a feed rate of 1 g-mole/sec, as found in Examples 1.2 and 2.2.

For a feed rate of only 1 g-mole per hour, the area in square feet is

$$A = \frac{1.16(10^6)}{3600(929)} = 0.35\,ft^2$$

In any event, the foregoing illustrates the obvious, that low membrane permeability can translate to significantly high equipment demands if high feed rates are involved, along with appreciable membrane thickness.

The corresponding spreadsheet-type calculations are shown in Appendix 3, which may be generalized.

3.6 ALTERNATE REPRESENTATION AND CALCULATION

The rate balances for a two-component system may be represented as

$$Vy_i = P_i(P_L x_i - P_V y_i)$$
$$Vy_j = P_j(P_L x_j - P_V y_j)$$

Therefore,

$$\frac{y_i}{y_j} = \frac{P_i}{P_j} \frac{\left(x_i - \dfrac{P_V}{P_L} y_i\right)}{\left(x_j - \dfrac{P_V}{P_L} y_j\right)}$$

or

$$\frac{y_i}{1 - y_i} = \frac{P_i}{P_j} \frac{\left(x_i - \dfrac{P_V}{P_L} y_i\right)}{(1 - x_i) - \dfrac{P_V}{P_L}(1 - y_i)}$$

Multiplying through and collecting terms,

$$0 = \left\{\frac{P_i}{P_j} \frac{P_V}{P_L} \cdot \frac{P_V}{P_L}\right\} y_i^2 + \left\{-1 + x_i - \frac{P_i}{P_j} x_i\right\} y_i + \left\{\frac{P_i}{P_j} x_i\right\}$$

This is a quadratic equation that may be solved for y_i, whereby y_i may be obtained in terms of x_i. The representation is not yet complete, however, for there coexists the material balance

$$F(x_F)_i = Vy_i + Lx_i$$

which, since $L = F - V$, may be expressed as the rearrangement

$$y_i = -\frac{1 - V/F}{V/F} x_i + \frac{1}{V/F}(x_F)_i$$

which is a straight line in y_i–x_i space for parametric values of the variable V or V/F. This relationship must be solved simultaneously with the previous result. The graphical intersection of the straight line for the material balance with a plot of the quadratic solution in y_i–x_i space yields the answer for a particular value of, say, V/F. This methodology is similar in principle to that shown by Hwang and Kammermeyer.[4]

An algebraic solution requires that the material balance be substituted into the quadratic form, which yields a new quadratic form, which in turn must be solved for a value of, say, y_i or x_i, depending on which variable is eliminated. The result in principle is identical to that of the previous sections and depends on the value specified for, say, V/F.

Note that the previously derived expression in the form

$$y_i = \frac{Vy_i}{V} = \frac{P_L P_i x_i - P_V P_i y_i}{P_L \sum P_i x_i - P_V \sum P_i y_i}$$

reduces to the following for a two-component system i and j:

$$y_i = \frac{P_i x_i - (P_V/P_L)P_i y_i}{[P_i x_i + P_j(1 - x_i)] - (P_V/P_L)[P_i y_i + P_j(1 - y_i)]}$$

or

$$y_i = \frac{(P_i/P_j)[x_i - (P_V/P_L)y_i]}{(P_i/P_j)[x_i - (P_V/P_L)y_i] + [(1 - x_i) - (P_V/P_L) + (P_V/P_L)y_i]}$$

This expression may be solved for y_i vs. x_i for parameters of (P_i/P_j) and (P_V/P_L). It is an approach advocated in a communication from Uzi Mann of the Department of Chemical Engineering at Texas Tech University.[5] The explicit solution for y_i turns out to be a quadratic equation, and graphical representations have been made by Mann in two-space for y_i vs. x_i using different parametric values of the pressure ratio $(P_V/P_L) = R$, with each representation involving a prescribed value for the permeability ratio $P_i/P_j = \alpha$.

For the record, performing the multiplication and collecting terms, the preceding equation converts to

$$(P_V/P_L)[1-(P_i/P_j)]y_i^2 + \{x_i[-1+(P_i/P_j)]+1-(P_V/P_L)$$

$$[1-(P_i/P_j)]\}y_i - x_i(P_i/P_j) = 0$$

or, on dividing through by a minus (P_V/P_L),

$$[(P_i/P_j)-1]y_i^2 + \left\{\frac{x_i[(P_i/P_j)-1]}{(P_V/P_L)} + [(P_i/P_j)-1]\right\}y_i + \frac{x_i(P_i/P_j)}{(P_V/P_L)} = 0$$

or

$$ay_i^2 + by_i + c = 0$$

where the quantities are defined by the substitutions. Solving the quadratic,

$$y_i = \frac{-b \pm \sqrt{b^2 - 4ac}}{2a}$$

where b and c are functions of x_i.

It may be observed that this can be written in the K-value form as

$$y_i = \left\{\frac{-b \pm \sqrt{b^2 - 4ac}}{2a} \frac{1}{x_i}\right\}x_i = K_i x_i$$

Here, K_i is a function of x_i rather than of the permeate flux rate V'' (or V''/F'') as previously derived.

Note: As a final notation, it has been assumed that the permeabilities are independent of composition and pressure and that the permeabilities determined for a single component has the same value in a mixture. That this is not necessarily the case has been determined by Lee and Minhas,[6] as indicated in Chapter 2, toward the end of Section 2.2, in the subsection Membrane Areas for Mixtures, where the permeabilities determined for pure gases proved much higher than the values occurring in mixtures.

This introduces the possibility of utilizing an efficiency rating for accommodating the discrepancy.

REFERENCES

1. Hoffman, E. J. *Azeotropic and Extractive Distillation.* New York: Wiley-Interscience, 1964; Huntington, NY: Krieger, 1977.
2. Hoffman, E. J. *Phase and Flow Behavior in Petroleum Production.* Laramie, WY: Energon, 1981.
3. Katz, D. L., D. Cornell, R. Kobayashi, F. H. Poettmann, J. A. Vary, J. R. Elenbaas, and C. F. Weinaug. *Handbook of Natural Gas Engineering.* New York: McGraw-Hill, 1959.
4. Hwang, S.-T., and K. Kammermeyer. *Membranes in Separations*, vol. VII of *Techniques of Chemistry*, ed. A. Weissberger, Chapter IV. New York: Wiley-Interscience, 1975.
5. Mann, U. "Simplified Performance Calculations of Gaseous Membrane Separators." Paper presented at the American Institute of Chemical Engineers Annual Meeting, San Francisco, November 1995.
6. Lee, S. Y., and B. S. Minhas. "Effect of Gas Composition and Pressure on Permeation through Cellulose Acetate Membranes." In *New Membrane Materials and Processes for Separation*, ed. Kamalesh K. Sirkar and Douglas R. Lloyd. AIChE Symposium Series, vol. 84, no. 261, 1987 AIChE Summer National Meeting, Minneapolis. New York: American Institute of Chemical Engineers, 1988.

4

Multistage Membrane Separations

A multistage membrane operation can be represented in similar fashion to a multistage, plate-to-plate, stepwise or stagewise distillation column or operation, although at first glance this may not appear readily evident. Moreover, by viewing and phrasing membrane separations in the terms used for distillation, the membrane separation derivations and calculations can be similarly systematized and similarly simplified. This requires the modification and utilization of the K-value concept as developed in the previous chapter, whereby the techniques for vapor-liquid flash vaporization calculations are adapted to single-stage membrane separations. The arguments are further pursued in this chapter, beginning with a review of multistage distillation.

4.1 MULTISTAGE DISTILLATION

A multistage distillation column can be represented schematically as diagrammed in Figure 4.1. The complete column is shown, with a condenser to produce external liquid reflux at the top and a reboiler at the bottom to produce an external recycle vapor phase. In the customary parlance, the section above the feed plate or feed point is called the *rectifying* (or *absorbing*) section, and the section below is called the *stripping* section. The usual symbolism is to denote the vapor phase or its molar flow rate by V or $\overline{V}$, and the liquid phase or its molar flow rate by L or $\overline{L}$, for the rectifying and stripping sections, respectively.

In sum, distillation is a countercurrent vapor-liquid operation with the external reflux of a condensed liquid phase at the top and the external recycle or reflux of reboiled or vaporized vapor at the bottom. This reflux or recycle feature produces a sharp separation between the two key components of the feed mixture, and the same applies to membrane operations. The key components are the two components of a mixture between which the separation is to be made.

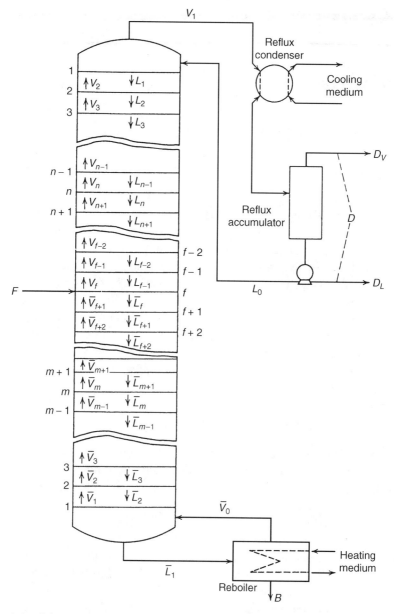

Figure 4.1 Schematic representation of a distillation column. (Source: E. J. Hoffman.[1])

The feed mixture or its molar flow rate is generally denoted by the symbol F. The upper or lighter, relatively more volatile product, called the *distillate product*, is denoted D. The lower or heavier, less volatile product, called the *bottoms* product, is denoted B.

Distillation calculations are commonly phrased in terms of an equilibrium stage or theoretical plate. By definition, the streams leaving each plate or stage are in equilibrium; that is, for example, the vapor stream V_n leaving the top of theoretical plate n is regarded as in equilibrium with the liquid stream L_n leaving the bottom of the same theoretical plate. Albeit this is not the case in actual practice, it permits calculation of the degree of separation versus the number of theoretical stages or plates used or vice versa, which can be compared or correlated to actual practice in terms of stage or plate efficiency.

The distillation operation is embodied in a vertical column to take advantage of the difference in density between the liquid and vapor phases. Thus, in stagewise distillation, the more-dense liquid phase L moves down vertically via downcomers at the side of the plate and next to the column wall. The liquid phase is thereby introduced at or on each plate and flows across the plate also by gravity; for example, by virtue of the hydraulic gradient that builds up in the downcomer. Meanwhile, the less-dense vapor phase V moves upward, passing up and through the liquid on the plate via holes or bubble caps (or nozzles) and overall moves countercurrently to the liquid phase, as represented schematically in Figure 4.1.

At the same time, a small pressure gradient forms up and down the column, with the highest pressure at the bottom. Normally in distillation practice, this pressure drop is minor compared to the working pressure, and the working pressure is assumed to be the same constant value up and down the column.

If viewed alone, without reflux or recycle, it may be noted that the top or rectifying section of a distillation column corresponds to simple vapor-liquid absorption; that is, to the absorption of a key component or components from a gas or vapor using a liquid absorbent or solvent. The bottom section similarly corresponds to simple liquid-vapor stripping; that is, to the stripping of the key component or components from a liquid using a gas or vapor phase.

Characteristically, neither absorption nor stripping produces a sharp separation unless, say, the solvent is highly specific to the selective absorption of the key component(s) or the vapor phase is highly specific to the selective removal of the key component(s). If reflux is used at the top of an absorber, for instance, there will be the tendency to produce the more-volatile component as the distillate product but with a poor or sloppy separation at the bottom. Conversely, if a reboiler is used at the bottom of a stripper, there will be the tendency to produce the less-volatile component at the bottom but with a poor or sloppy separation at the top.

Furthermore, a distillation operation can also be described differentially in terms of a continuum as presented, for instance, by Hoffman.[1] This representation is more suitable for, say, packed columns vis-à-vis plate columns, although the information is sometimes interchangeable; for example, in terms of the height equivalent of a theoretical plate (HETP). Other concepts used include what is referred to as the height of a transfer unit (HTU) and the number of transfer units (NTUs).

(The line of argument pursued here can be extended to liquid-liquid extraction operations. If there is an appreciable density difference between the two immiscible liquid phases, then a vertical column can be used. Otherwise, interstage pumping of the liquid phases is necessary, and the stages can be juxtaposed in any manner. Extension can be made to liquid-solid separations, even to vapor-solid separations, although the representations are increasingly less satisfactory.)

Calculation Methods for Binary Distillations

In brief review, the principal methods used for binary distillation calculations are graphical, known as the McCabe-Thiele method and the Ponchon-Savarit method.[1,2] The McCabe-Thiele method assumes constant molal overflow utilizing a vapor-liquid y-x diagram and operating lines for the rectifying and stripping sections. These operating lines are straight lines with slopes L/V and $\overline{V}/\overline{L}$, respectively. The number of stages is stepped off between the y-x equilibrium curve and the operating lines, which is an application of the material balances for the successive stages.

(Interestingly, these graphical representations can be extended to three components in the plane, with the third mole fraction a dependent variable, and to four components in three-dimensional space, as presented by Hoffman.[1])

In the Ponchon-Savarit method, an enthalpy-composition (H-x) diagram is utilized for the mixture (but which is known for relatively few two-component systems, ethanol and water being the preeminent example studied). This is accompanied by a y-x diagram that establishes the behavior of the vapor-liquid tie-lines at equilibrium on the enthalpy-composition diagram. The simultaneous consideration of the enthalpy balance and material balance at each stage leads to the "delta point" concept, whereby all such combined balances converge at a single point in enthalpy-composition space. There is a common delta point for the rectifying section and another for the stripping section.

A feature of distillation is that the temperature changes from stage to stage, as do the equilibrium-stage compositions. (For multicomponent systems, the flash-vaporization type of calculation for the equilibrium condition introduces an element of trial and error at each stage.) The column operating pressure is generally assumed to remain constant, albeit in practice there is a slight pressure increase from top to bottom.

An enumeration of the variables and equations involved establishes the degrees of freedom, by difference.[1] That is, subtracting the number of independent equations from the number of independent variables leaves the number of degrees of freedom. This denotes the number of variables that must be assigned values to effect solutions.

When stage-to-stage enthalpy and material balances are involved, the number of degrees of freedom is 3, whether we are speaking of two components or of a multicomponent separation.[1] If only the material balances are involved, then the number of degrees of freedom is 4. It must be kept in mind, however, that we are speaking of an integral number of stages, that is, an integral number of equations.

The degrees of freedom for the McCabe-Thiele method can stand further examination, therefore, since only material balances are involved, as is the case here for membrane separations. As ordinarily applied, a constant operating pressure is specified, which affixes the y-x diagram. The separation is then generally specified, that is, as $(x_D)_i$ for the one component in the distillate product D, and $(x_B)_i$ for, say, the same component of the bottoms product B (the other component mole fractions are dependent variables, since only two components are involved). The operating line in the rectifying section may be affixed by specifying the operating slope L/V, with its terminus at $(x_D)_i$.

Note that the 4 degrees of freedom have now been used up. Properly speaking, the graphical solution should then entail trial-and-error procedures involving an integral number of steps, which would be required to proceed from $(x_D)_i$ to $(x_B)_i$. The latter point is the terminus for the operating line for the stripping section, which has a slope $\overline{V}/\overline{L}$, yet to be determined by the trial-and-error procedure. The operating lines can be assumed to intersect at the feed plate or feed stage, whereby the feed plate material balance would determine the partitioning of the feedstream F between introduction as a liquid and as a vapor. This material balance can, in fact, be plotted graphically as a straight line, with the slope related to the fraction of liquid X and fraction of vapor $(1 - X)$.

In practice, however, a different tack is generally taken. The operating pressure is again routinely specified, which affixes the y-x vapor-liquid equilibrium curve. In turn, the separations $(x_D)_i$ and $(x_B)_i$ are also specified, followed by the specification of an internal reflux ratio L/V (or external reflux ratio L/D). Note that the 4 degrees of freedom have been assigned values, as before.

Instead of proceeding with a trial-and-error solution for the internal reboil ratio $\overline{V}/\overline{L}$ (or the external reboil ratio $\overline{V}/B$), the reboil ratio $\overline{V}/\overline{L}$ is instead specified, which affixes the operating line for the stripping section. Alternately and equivalently, the value for the liquid molar feed fraction X can instead be specified, which also affixes the operating line for the stripping section.

Now, 5 degrees of freedom have been specified, which overspecifies the system of variables and equations. However, this overspecification is accommodated merely by stepping off a fractional number of stages—which, for all practical purposes, is no doubt close enough—especially when dealing with the theoretical stage or equilibrium stage concept, since this is an approximation in itself. This is related to the real world by introducing the idea of stage efficiencies or other devices to correlate theory with experiment. In turn, stage efficiencies require their own correlations.[1]

Integral Number of Stages

In stage-to-stage or plate-to-plate distillation calculations, the method of calculation necessarily involves an integral number of stages or steps. This is because the calculations are performed analytically, equation by equation and step by step.

The set of calculations can be started at either end of the column and proceed toward the other, introducing the feedstream at some intermediate step, either as a (saturated) liquid, (saturated) vapor, or some partitioned combination (or even as a supercooled liquid or superheated vapor). Alternately, the calculations may start at the (partitioned) feedstream location, and proceed toward both ends.

If enthalpy balances are not involved, only material balances and phase equilibria, then constant molal overflow is assumed and in principle 4 degrees of freedom exist. The assignment of a uniform operating pressure uses 1 degree of freedom. In starting the calculation at, say, the distillate product end, with an assigned composition $(x_D)_i$ and reflux ratio L/D or L/V and assuming a partitioning (X) of the feed, the 4 degrees of freedom are used up. The reboil ratio $\overline{V}/B$ or $\overline{V}/\overline{L}$ becomes dependent by the assignment of X. (Alternately, the reboil ratio could be assigned, then X becomes dependent.) Furthermore, after an unspecified integral number of calculational steps, the composition so determined at the other end must satisfy the overall material balance with the feedstream—the overall calculational sequence can therefore be viewed as trial and error—and introduces the matter of "meshing," or comparing one calculated value (or set of calculated values) against another (not to mention that the phase equilibrium and material balance determination at each plate in general requires a trial-and-error procedure). Otherwise, of course, the calculation sequence may in similar fashion start with the bottoms product and continue upward toward the distillate product.

As still another possibility, the calculations may start at both the distillate product end and bottoms product end and meet (more or less) at the feedstream location. The assignment of an operating pressure uses 1 degree of freedom, and the assignment of $(x_D)_i$ and $(x_B)_i$ uses 2 more (and by the overall material balances sets the B/D ratio). Specifying either the reflux ratio or reboil ratio uses the fourth. Trial and error is used to determine a feedstream partitioning X such that the calculation meet at the feed location for an integral number of steps in each section. The option, of course, is to assume yet another degree of freedom (such as the reboil ratio, if the reflux ratio is already specified) and allow for the relaxation of exactness.

It may be emphasized for the preceding that no restrictions are placed on the feedstream composition or its partitioned phases, identifying with or being equal to stage compositions at the feedstream location.

If the calculation is started at the feed location, however, some sort of accommodation must be made with the feedstream and its composition. Therefore, an assignment can be made about the partitioning of the feedstream between liquid and vapor phases. Moreover, there is the matter of whether these compositions are to identify with the phase compositions at the feed stage. This agreement may in fact be made a contingency and establishes the composition or compositions at the feed stage. In other words, the feed-stage vapor-liquid mole fractions are made equal to the feedstream-partitioned vapor-liquid mole fractions. This qualification, in effect, utilizes 3 more degrees of freedom, including establishing a value for X. Hence, the requisite 4 degrees of freedom already are utilized. Any further assignments overspecify the system. Nevertheless, we proceed.

The calculation then becomes a matter of assuming values for L/D or L/V or for $\overline{V}/B$ or $\overline{V}/\overline{L}$. (The values of L/V and $\overline{V}/\overline{L}$ are related, along with X, by a material balance at the feed stage.) These values can then be used to calculate $(x_D)_i$ and $(x_B)_i$. The latter calculated values must then agree with the overall material balances with the feedstream F:

$$F = D + B$$
$$F(x_F)_i = D(x_D)_i + B(x_B)_i$$

whereby

$$\frac{B}{D} = \frac{(x_D)_i - (x_F)_i}{(x_F)_i - (x_B)_i}$$

In turn, it so happens that B/D can be related to X, L/V, and $\overline{V}/\overline{L}$, completing the circle. This is demonstrated subsequently as applied to membrane separations. Moreover, the calculation is properly carried out in an integral number of steps (equations).

With the degrees of freedom thus overspecified, there is the inference that an exact solution cannot be attained, no matter how many or what steps are utilized. The denouement then becomes a matter of meshing or approximation.

Considering that approximation is the name of the game, anyway, for convenience sake, the calculations for membrane separations proceed from a feed location in both directions; that is, proceed upward in a rectifying section and downward in a stripping section, with the separations so attained depending on the integral number of stages or membrane cells (equations) to be used for each section. This mode is particularly well suited to the absorption/factor and stripping/factor concept, where the K-values may be assumed constant.

Multicomponent Stagewise Distillations

Assorted computer programs for distillation are available, which spit out the numbers, and for sharp separations between the two key components. For the components outside, or well outside, the volatility range of the key components, it does not much matter anyway. The more-volatile components mainly go to the distillate product, the less volatile to the bottoms product, and any scheme of proration works.

The difficulty with a completely rigorous treatment is that the degrees of freedom do not increase with the number of components. And the degree of trial and error required increases with the number of components.

4.2   THE ANALOGY

A multistage membrane process can be perceived as represented successively in Figures 4.2–4.5. Figure 4.2 would correspond to the top or rectifying section of a distillation column, Figure 4.3 to the middle or feed section, and Figure 4.4 to the bottom or stripping section. Figure 4.5 is a juxtaposition of Figures 4.2–4.4; that is, is a juxtaposition of the three sections.

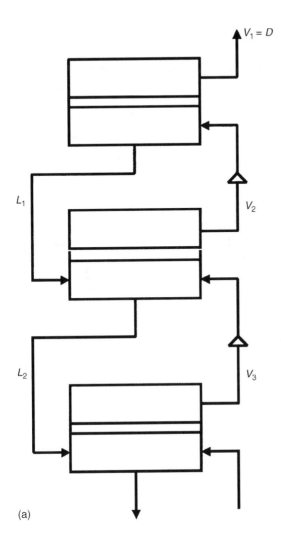

$V_1 = D$

L_1

V_2

L_2

V_3

(a)

Figure 4.2 Schematic representation of a membrane operation corresponding to the top section or rectifying section of a distillation column: (a) without external recycle or reflux; (b) with external recycle or reflux.

The double line in each cell denotes the membrane proper and corresponds to a plate or stage in distillation, the latter as represented by a single horizontal line in Figure 4.1. It may be observed as a distinguishing feature that the phase designated L is introduced immediately below the membrane, on the high-pressure or reject side of the membrane and, in this case, is a gaseous phase, as is phase V, whereas in distillation practice, the liquid phase L is introduced at or onto the plate.

For purposes of further clarification and in the juxtaposition illustrated in the figures, the reject from the membrane cell above becomes part of the feed to the reject side of the cell below, along with the stream V

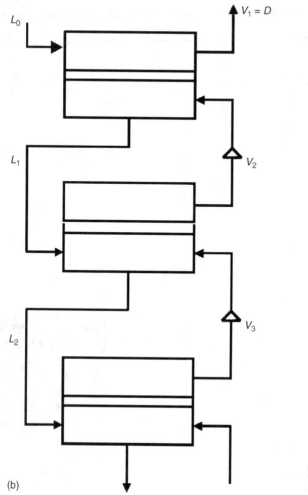

(b)

Figure 4.2
(continued)

from the next cell further below. The juxtaposition for the multistage membrane assembly or unit as a whole (Figure 4.5) may alternately be referred to as a *cascade operation*, or *arrangement*, using an intermediate feed location for the feedstream designated *F*.

The action is such that the more-permeable component(s) move progressively toward the top of the membrane cell assembly, as per the vertical juxtaposition used in the figures, and the less-permeable component(s) move toward the bottom. This is entirely analogous to distillation, where the more-volatile component(s) move progressively toward the top of the column and the less volatile component(s) move toward the bottom of the column.

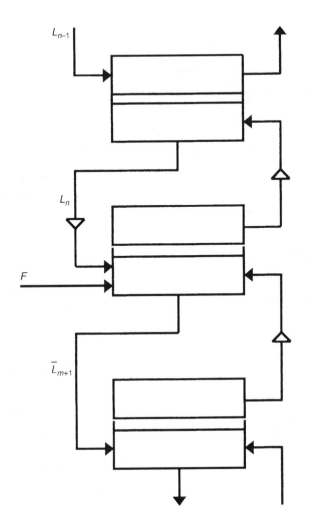

Figure 4.3 Schematic representation of a membrane operation corresponding to the middle section or feed section of a distillation column.

Flow streams designated L here represent the high-pressure side (P_L) of each membrane cell, and the exiting flow streams designated V represent the low-pressure side (P_V). The low-pressure side corresponds to the permeate side of each membrane, the high-pressure side to the feed/reject side.

It is necessary that compression occur between the cells or stages to maintain the necessary or specified pressure difference across each successive membrane and circulation between successive stages. Moreover, this compression can be used to maintain more or less uniform flow rates for the permeate and reject phases between successive stages.

Alternately or in addition to the compression of the gaseous streams V between stages, the gaseous streams L may be compressed between

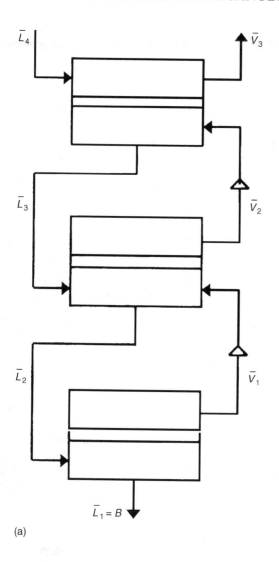

Figure 4.4 Schematic representation of a membrane operation corresponding to the bottom section or stripping section of a distillation column: (a) without external recycle or reflux; (b) with external recycle or reflux (or reboil).

stages. If only the streams designated V are compressed, the pressures on the reject side progressively increase, going upward in the presentation. This ensures flow of the reject/feed from cell to cell, which may be regulated by a valve. If the streams designated L are also compressed, only to a smaller degree, then the pressure sequence upward or downward can be made arbitrary.

In fact, for the convenient purposes here, the pressures on the reject side of the membranes can be assumed to take on the uniform constant value P_L, and the pressures on the permeate side can be assumed to take on the uniform constant value P_V.

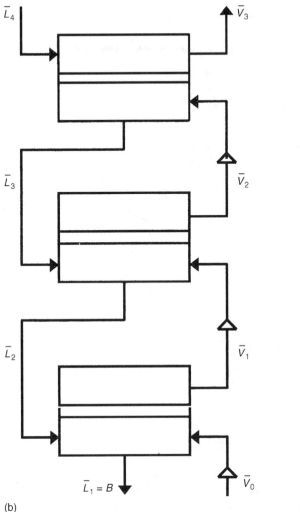

Figure 4.4
(continued)

(b)

The layout is thus similar to a distillation column but without introducing an external reflux or recycle at the top and without reboil at the bottom. Alternately, the top cell can be viewed as initiating the reflux stream L, and the bottom cell as initiating the reboil stream $\overline{V}$. In distillation column parlance, the top membrane cell then corresponds to the reflux condenser, and the bottom membrane cell corresponds to the reboiler. That is, in distillation operations, a partial reflux condenser and accumulator can be perceived as an extra stage (the distillate product is recovered as a vapor), and the reboiler (called a *partial reboiler*) as an extra stage if the bottoms product is recovered as a liquid.

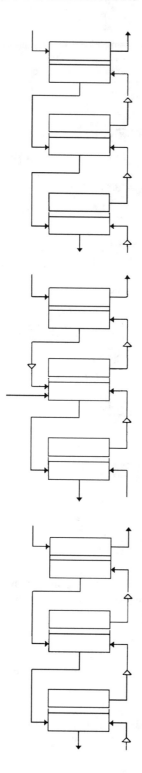

Figure 4.5 Combined multistage membrane operations.

It may be noted that in distillation, a partial condenser will produce a change in composition, whereas a total condenser does not produce a change in composition. Accordingly, the recyle or reflux embodiment of Figure 4.2(a) corresponds to total condensation; that is, there is no composition change.

Also in distillation, a partial reboiler at the bottom of the column produces a change in composition and can be viewed as an extra stage, whereas a total reboiler does not produce a change in composition (all the recycled liquid is vaporized before returning to the column). However, in the embodiment of Figure 4.4(b), the analogy cannot apply, since the recycled stream $\overline{V}_0$ not only can have the same composition but is not ordinarily introduced into the opposite or permeate side of the membrane cell—compromising the composition of $\overline{V}_1$—but into the high-pressure reject side. In other words, it is used to merely make another "pass" on the reject side, which as noted elsewhere, merely increases the size of the membrane cell area but theoretically does not enhance the degree of separation at the cell, assuming perfect mixing. However, the introduction of the recycle stream enlarges the cell permeate rates up through the stripping section, at the same time making the assumption of a constant permeate rate (and reject rate) up through the unit a more likely proposition for calculation purposes.

As in distillation, sidestreams can be introduced or sidestream products can be withdrawn at different points or stages. In distillation practice at least, it is preferable that the sidestreams introduced be approximately of the same composition as at the point or stage introduced. Again by analogy, the membrane section "above" the feed location may be referred to as the *rectifying* section, and the section "below" as the *stripping* section or vice versa. The former is the accepted convention in distillation practice, where the more-volatile component(s) move toward the top of the column, the less-volatile toward the bottom.

Although the cells may be laid out horizontally side by side or in other juxtapositions, the convention is adopted here, as previously observed, that the more-permeable component(s) preferentially move toward the "top" whereas the less-permeable component(s) preferentially move toward the "bottom."

For convenience in the ensuing calculations, we restate that the high-side pressure P_L has the same constant value throughout, similarly for the low-side pressure P_V. Furthermore, as noted, interstage compression is required to convert the pressure on the low-pressure side to the pressure on the high-pressure side of the next succeeding cell. As indicated, the high-pressure side of a cell corresponds to the reject side, whereas the low-pressure side corresponds to permeate side.

In further explanation and reiteration, the permeate stream leaving a cell becomes a feedstream to the next adjacent cell (here, upward). The reject from this next cell is returned as a feedstream of the first-mentioned or anterior cell, and so forth.

Moreover, the recycle rate between cells is a process variable or parameter to be set and accommodated to the membrane size and permeability and to the pressure drop across the membrane. Affixing the compression rate for the permeate phases tends to affix the corresponding reject rate or vice versa.

And so on, all up and down the line, whereby a condition of essentially constant permeate rates may be maintained from stage to stage and, at the same time, essentially constant reject rates. This simplification is akin to constant molal overflow as assumed in distillation calculations and makes multistage membrane calculations manageable.

Efficiency

In utilizing equilibrium stage distillation calculations, it is the common practice to introduce stage efficiencies, which are based on actual operating results. These efficiencies are generally less than 100%.

The same is true in utilizing membrane calculations. As indicated by S. Y. Lee and B. S. Minhas,[3] the observed permeabilities for the components of a mixture may be markedly less than that measured for the individual pure components. This indicates that there is a role for efficiency ratings, which may be stagewise or pointwise (as in the case of differential permeation) or an overall figure may be used.

4.3 GRAPHICAL REPRESENTATION OF BINARY MEMBRANE CALCULATIONS

What is often called the *McCabe-Thiele method*[1,2,4] for binary distillation calculations deploys a y-x (or $\bar{y}$-$\bar{x}$) diagram, say, for the more-volatile component, here designated the more-permeable component i. This furnishes substantiation that a separation can indeed be attained by the use of recycle or reflux in a multistage or cascade operation.

Consider, therefore, Figure 4.6, which denotes a graphical membrane calculation based on the McCabe-Thiele method for distillation. The ordinate, y here, denotes the composition of the permeate phase(s) V. The abscissa, x here, denotes the composition of the reject phase(s) L. Constant values of V and L are assumed throughout, equivalent to a condition of constant molal overflow. To represent the equations, a continuum is

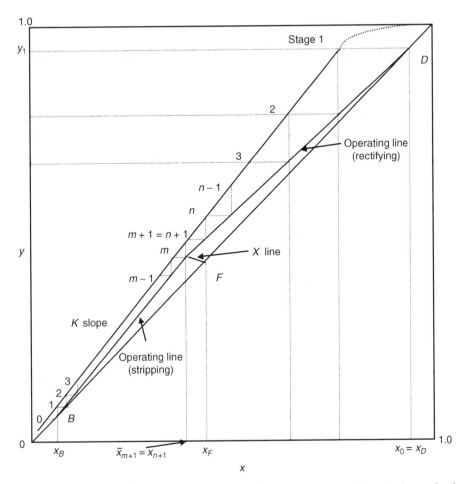

Figure 4.6 Graphical schematic representation for the McCabe-Thiele method as applied to membrane separations. The value of K is to remain essentially constant over the domain of application and, for the more-permeable component i, has a slope greater than unity. As the pure component i is approached, the behavior of K in principle terminates at $y = x = 1$.

assumed, albeit the equations actually represent step functions, with a point for each membrane cell.

A 45° diagonal is drawn across the figure, where $y = x$, and on this diagonal, the points are schematically located designating the compositions x_F, x_D, and x_B for the more-permeable component i.

Operating lines and the intermediate behavior of the X locus or X line (or q line) are sketched in for a partitioning of the feedstream. Moreover, a few stage calculations are schematically shown at the more-permeable product end and for agreement between the feed and membrane

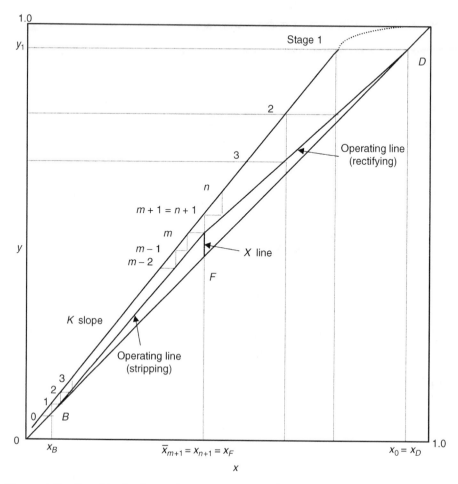

Figure 4.7 Graphical schematic representation for the McCabe-Thiele method as applied to membrane separations, where $X = 1$.

cell compositions at the feedstream location. Whereas Figure 4.6 assumes a partitioning of the feedstream between the permeate and reject streams at the feed location, in Figure 4.7 all of the feedstream is assigned to the reject phase, as indicated by the vertical q-line, signifying that $X = 1$. In Figure 4.8, all of the feedstream is assigned to the permeate phase, as indicated by the horizontal q-line, signifying that $X = 0$.

K-Value Behavior

A distinction is made, however, in that, in distillation, the temperature and phase equilibrium compositions vary. For membrane calculations, on the other hand, the relation between the reject and permeate

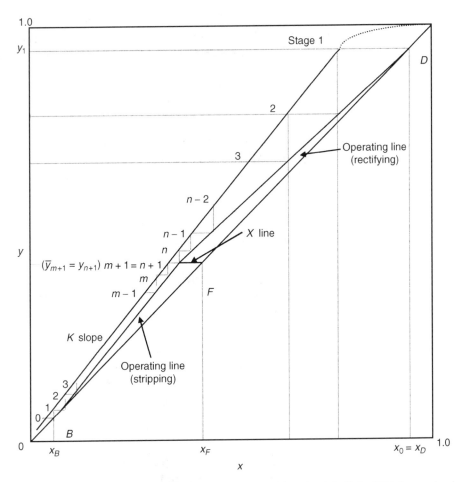

Figure 4.8 Graphical schematic representation for the McCabe-Thiele method as applied to membrane separations, where $X = 0$.

phases is regarded as uniform; that is, at each stage n (or m) the K-value for component i is the same constant value, represented by

$$K_i = \frac{P_i P_L}{V'' + P_i P_V}$$

where V'' denotes the permeate flux, to be determined as derived and utilized in Chapter 3. This has far-reaching consequences, in that analytical rather than graphical methods can be used in the calculations, as is subsequently developed.

The "equilibrium" curve, or K-value, is for the most part represented by a straight line with slope K (or K_i). Since this pertains to the more-permeable component, the slope is greater than unity. Furthermore, the K-value line lies above the 45° diagonal, albeit the actual determination is to a certain extent arbitrary, as per Example 3.1 of Chapter 3; that is, is it determined from a bubble-point, dew-point, or in-between type calculation on the feedstream composition?

Moreover, for a pure component (for both i and j), the limiting value for $K = y/x$ should be unity. Accordingly, dashed or dotted lines are sketched in to accommodate this boundary condition at the upper end. Moreover, as $y = x = 1$ is approached, the slope dy/dx does not remain equal to the presumed constant value for K, since K then has to vary.

The exact behavior for component i at the far lower end is assumed mostly unknown. However, the K-value for component j necessarily approaches unity.

In other words, the behavior of the straight line here representing a constant value for K_i does not terminate exactly at the origin, with a gap left for the unknown. In the drawings, the behavior is denoted more or less schematically, without continuing the representation down to the origin. However, the most probable idealized behavior would seem to be that where the straight line continues down to the origin, with the annotation that, when $y = x = 0$, the ratio y/x becomes indeterminate.

Note that, for the pure component i, when $K_i = 1$, it follows that

$$V'' = P_i(P_L - P_V)$$

A corresponding limiting condition occurs for pure component j.

(Vapor-liquid equilibria pose a similar problem; that is, the behavior of the equilibrium vaporization ratio K as experimentally determined for a mid-range of temperature, pressure, and composition does not apply for the pure component nor at supercritical or near-supercritical conditions, such as for the light hydrocarbons, as per Katz et al.[5] As an approximation, however, the equilibrium vaporization ratio may be defined by Raoult's law, that $K = vp/P$; that is, the vapor pressure of the pure component divided by the total pressure. In turn, the mixture obeys Dalton's law of partial pressures. By these accommodations, the boundary condition for the pure component can be met: $K = 1$.)

Note that, in the representations of Figures 4.6–4.8, the K slope is greater than unity, so that the y ordinate and x abscissa refer to the more-permeable component. Note furthermore that, for this juxtaposition, the same degree of separation requires more stages or cells in the stripping section than for the rectifying section.

Operating Lines

In Figure 4.6, operating lines are posted for the rectifying sections and stripping sections, terminating at points on the diagonal respectively marked D and B, for the more-permeable and less-permeable products. These straight lines are determined from the material balances as a continuum.

For the rectifying section,

$$V = L + D$$

$$Vy = Lx + Dx_D$$

whereby

$$y = \frac{L}{V}x + \frac{D}{V}xD$$

It may be readily observed that this operating line terminates on the diagonal at the point denoted D, where $y = x = x_D$.

For the stripping section,

$$\overline{L} = \overline{V} + B$$

$$\overline{L}\,\overline{x} = \overline{V}\,\overline{y} + Bx_B$$

whereby

$$\overline{y} = \frac{\overline{L}}{\overline{V}}\overline{x} - \frac{B}{\overline{V}}x_B$$

The overbars or overlines are used to designate entities in the stripping section. Note that this operating line terminates at the point designated B, where $\overline{y} = \overline{x} = x_B$.

Feedstream Partitioning

At the feedstream location or feed cell, the stream material balance is

$$F + L + \overline{V} = \overline{L} + V \quad \text{or} \quad \overline{V} = \overline{L} + V - L - F$$

In turn, let

$$\overline{L} = L + XF \quad \text{and} \quad V = \overline{V} + (1 - X)F \quad \text{or} \quad \overline{V} = V - (1 - X)F$$

where X denotes the (molar) fraction of the feedstream introduced into the reject side of the membrane cell at the feed location, and $(1 - X)$

denotes the (molar) fraction of the feedstream introduced into the permeate side.

(In Brown et al.[2] and most other references, q is used to denote this fraction. However, since the symbol applies to a feedstream of combined saturated liquid and vapor phases—and even to supercooled liquid and superheated vapor as the feedstream—and as applied to distillation, the symbol X is used here instead for membrane separations. It may be added that the following derivations are circuitous and not the only way to represent the effect of feedstream partitioning but are the generally accepted way and perhaps the most convenient to use.)

Therefore,

$$\overline{V} = V - (1 - X)F = L + D - (1 - X)F = L + (F - B) - (1 - X)F = L - B + XF$$

Substituting for $\overline{L}$ and $\overline{V}$ into the foregoing material balance representing the operating line for the stripping section,

$$\overline{y} = \frac{L + XF}{(L - B) + XF}\,\overline{x} - \frac{B}{(L - B) + XF}\,x_B$$

Multiplying through and collecting terms,

$$[(L - B) + XF]\overline{y} = [L + XF]\,\overline{x} - Bx_B$$

$$L(\overline{y} - \overline{x}) = XF(\overline{x} - \overline{y}) + B(\overline{y} - x_B)$$

For the rectifying section, since

$$y = \frac{L}{V}x + \frac{D}{V}x_D = \frac{L}{L + D}x + \frac{D}{L + D}x_D$$

then on multiplying through and collecting terms,

$$L(y - x) = D(x_D - y)$$

For the point of intersection of the two operating curves, $\overline{y} = y$ and $\overline{x} = x$. Therefore,

$$D(x_D - y) = XF(x - y) = B(y - x_B)$$

From the overall material balances,

$$B = F - D$$

$$Bx_B = Fx_F - Dx_D$$

A substitution yields

$$D(x_D - y) = XF(x - y) + (F - D)y - Fx_F + DxD$$

Collecting and canceling terms,

$$0 = XFx - XFy + Fy - Fx_F$$

or

$$y(X - 1) = Xx - x_F$$

$$y = \frac{X}{X-1}x - \frac{1}{X-1}x_F = -\frac{X}{1-X}x + \frac{1}{1-X}x_F$$

Therefore, the locus for the intersection of the operating lines is a straight line with a negative slope of $X/(1 - X)$ and intersects the point designated F on the 45° diagonal, where $y = x = x_F$. When $X = 1$, the line is vertical (all the feedstream goes to the reject side), and when $X = 0$, the line is horizontal (all the feedstream goes to the permeate side).

Number of Stages

The number of membrane stages or cells is determined from a stepwise procedure, starting, say, from the more-permeable product D. It can be assumed that the recycle or reflux ratios for the rectifying and stripping sections have been specified, and the feedstream partitioning X, if any. The first few steps are illustrated in Figures 4.6 through 4.8.

In fact, various combinations of variables can be specified, some of which are independent and others dependent. These variously include the recycle or reflux ratios in each section, say, L/V and $\overline{L}/\overline{V}$ or $\overline{V}/\overline{L}$, plus the feedstream partitioning X, and the compositions x_F, x_D, and x_B. Alternately, the stepwise procedure may originate at the less-permeable product B, the feedstream location, or at the feed cell and proceed in both directions.

Whichever starting point is used, the number of stages or cells most surely is a fractional number; that is to say, starting at D or x_D, the stepwise procedure does not end exactly at the point designated B or x_B. This is a signal that the degrees of freedom have been overspecified. Otherwise, trial-and-error procedures must be instituted—maybe even double or triple trial and error, depending on the makeup of the equations—if an integral number of steps is to be achieved.

There is a further question of whether there should also be an integral number of stages or cells in each section, that is, in both the rectifying and stripping sections. In a McCabe-Thiele type of plot, the composition(s) at the feed stage or cell is represented by a point on the locus for the partitioning of the feedstream, the "X locus." This point denotes where the permeate composition becomes the same for both the rectifying and stripping sections, and likewise for the reject composition. In other words, cell $n + 1$ becomes identical with cell $m + 1$ (and the feed location, cell f). The permeate composition and reject compositions for this particular cell are defined by the intersection. (A further discussion is furnished in Example 4.1).

This juxtaposition, however, necessarily excludes the point where the X locus intersects the K-value representation (or equilibrium curve), since this would lead to a condition of minimum reflux and infinite stages, as described in the next subsection.

In summation, rigorously speaking, we are speaking of an integral number of equations, although it probably does not matter too much, everything else considered.

Stepwise Graphical Calculations

A brief iteration of the stepwise process may be made, say, as per Figure 4.6, where the graphical representations are kept sparse to avoid unnecessary clutter. The calculation may be perceived as starting at D, the initial point on the 45° line, where $y_1 = x_0 = x_D$.

The corresponding point on the K-value line (the line designated K slope) is where y_1 is in "equilibrium" with x_1; that is, the point (y_1, x_1).

In turn, this value of x_1 next permits the calculation of y_2 via the pointwise relationship

$$y_2 = \frac{L}{V} x_1 + \frac{D}{V} x_D$$

This relationship, however, merely denotes a point on the operating line for the rectifying section and is obtained by dropping down vertically from the point (y_1, x_1). In other words, the operating line is merely the locus connecting the successive points $(y_2, x_1), (y_3, x_2), (y_4, x_3)$, and so forth. In turn, knowing y_2, the value x_2 follows from the K line and so forth.

Alternately, the graphical calculation can start at the feedstream location and proceed toward D. In Figure 4.6, the point on the K-value or "equilibrium" line denoted by $m + 1 = n + 1$ signals that, at the feed location,

$$\bar{y}_{m+1} = y_{n+1} \quad \text{and} \quad \bar{x}_{m+1} = x_{n+1}$$

Furthermore, that

$$\bar{y}_{m+1} \text{ is in "equilibrium" with } \bar{x}_{m+1}$$

$$y_{n+1} \text{ is in "equilibrium" with } x_{n+1}$$

However, the representation shows that the partitioned feedstream introduced does not have the same composition as either the reject or permeate sides of the cell. This introduces an unnecessary complication, whereby analytical methods require an adjustment or accommodation with the reject and permeate compositions at the feedstream cell.

(These adjustments or accommodations can be made by material balances on the streams involved. In distillation, on the other hand, if the feedstream is assumed to be a mixture of saturated liquid and saturated vapor at equilibrium, then no such accommodation is necessary. The saturated liquid phase can go to the stage or plate liquid phase and be assumed of the same composition, and the saturated vapor phase can go to the stage or plate vapor phase and be assumed of the same composition.)

As the limiting case, in Figure 4.7, all the feedstream is introduced on the reject side and assumed to have the same composition as the reject stream. In Figure 4.8, all the feedstream is introduced on the permeate side and assumed to have the same composition as the permeate stream The former is the preferred embodiment here, in that the permeate load stays the same in both the rectifying and stripping sections. Note that the positioning of the operating lines changes for the different allocations of the feedstream.

If calculations are to start at the feedstream location, proceeding toward D in the rectifying section, then the corresponding calculation in the stripping section proceeds toward B. An abbreviated stepwise sequence is indicated in the figures.

Conversely, the calculation for the stripping section can start at point B, where

$$x_B = \bar{x}_1 = \bar{y}_0$$

Knowing $\bar{x}_1$ on the "equilibrium" curve gives $\bar{y}_1$. In turn, knowing $\bar{y}_1$ permits the calculation of $\bar{x}_2$ via the equation

$$y_1 = \frac{\overline{L}}{\overline{V}} x_2 - \frac{B}{\overline{V}} x_B$$

This determination may be perceived as merely a point on the operating line for the stripping section and so forth. The sequence is briefly indicated in the figures.

Comments

A few observations, therefore, emanate from a visual inspection of Figures 4.6–4.8. As far as the total number of stages is concerned, it makes very little difference where the calculation is started and ended, albeit obtaining a trial-and-error mesh for a near-integral number of stages can be a problem, one sometimes or oftentimes better ignored.

The important thing is that using recycle (or reflux) of the more-permeable product and recycle (or reflux) of the less-permeable product can induce sharper separations, just as in distillation. Furthermore, as far as the total number of stages is concerned, it makes very little difference as to whether the feedstream is introduced at the reject or permeate side of the cell at the feed location and whether or not it is partitioned. Not only this, it makes very little difference whether the feedstream composition coincides with the reject or permeate compositions.

The overriding consideration is that of convenience; in other words, what simplifies or eases the modus operandi for the calculations. And, as is further explained and demonstrated for analytical calculations, it is by far preferable to start calculations at the feedstream location, assuming that the feedstream composition, and preferably the reject composition, are identical. In this way, the overall material balance is automatically satisfied, even for an integral number of stages or cells in each section, which, not so incidentally, is a fundamental feature for the analytical calculation.

Limiting Conditions

The two main limiting conditions are total reflux (or recycle) and minimum reflux (or recycle). In total reflux, $L/V = \overline{L/V} = 1$, and both the operating lines coincide with the 45° diagonal. This condition gives the fewest number of stages for a given separation. The stages or cells are stepped off between the K line and the 45° diagonal. Again, the question arises of whether an integral number of stages should be pursued. Moreover, no finite product streams are obtained.

Minimum reflux, by definition, is that condition whereby a zone of an infinite number of stages exists "immediately" on each side of the feed location.[1] In the McCabe-Thiele method of graphical representation and calculation, both operating lines intersect with the locus of the feed partitioning, all at a point on the K-value plot (or equilibrium curve). It is therefore impossible to calculate away from this point of intersection, in either direction.

Analytical Methods of Analysis and Calculation

However, analytical procedures based on the absorption/stripping factor concept permit an integral number of stages or cells to be specified beforehand for each section. Furthermore, the calculations terminate at points x_D and x_B, so determined after the fact by the calculation procedures and, in the process, automatically satisfying the material balances. These analytical methods are next developed in detail.

4.4 RECTIFYING SECTION

In what is called the *rectifying section*, the stagewise material balances for a component i are as follows:

$$V_{n+1}y_{n+1} - L_n x_n = Dx_D$$

where, for simplicity, the component subscripts have been dropped.

If the V_{n+1} or V_n and the L_n are made essentially constant up and down the column or unit (a condition called *constant molal overflow* and *underflow*), then

$$Vy_{n+1} - Lx_n = Dx_D$$

or

$$y_{n+1} = \frac{L}{V}x_n + \frac{D}{V}x_D$$

This is represented by a straight line in y-x space, which takes on values at the successive points y_{n+1} and x_n. It is therefore a step function or difference equation. The slope L/V is less than unity, since, by the total stream balances,

$$V_{n+1} - L_n = D \qquad \text{or} \qquad V - L = D$$

The membrane rate balance across the nth membrane for each component i is of the form

$$Vy_n = P_i(P_L x_n - P_V y_n)$$

This relationship assumes perfect mixing, whereby the composition x_n is uniform across the high-pressure side of the nth cell, and the composition y_n is uniform across the low-pressure side of the cell. In other words, the composition of a stream leaving a cell is regarded as the composition within the totality of that side of the cell. This indeed is the meaning of the term *perfect mixing*.

Rearranging, the K-value form is attained:

$$y_n = \frac{P_i P_L}{V'' + P_i P_V} = K x_n$$

where V'' denotes the permeate molar flux and can be made numerically equal to the molar rate V if so desired, which calls for adjusting the other stream rates. The subscript n is dropped from the value of K since here K (or K_i) remains constant from cell to cell, for each component. For convenience in the notation, it is understood that K pertains to any component i.

The aforementioned material balance is a step equation or difference equation that takes on values at positions 1, 2, 3,..., n. It plots as a straight line in y-x space with slope K and intercept at the origin.

Note that, if $K > 1$, then $y_n > x_n$ for some component i. The component is preferentially passed through the membrane from the high-pressure to the low-pressure side; that is, the membrane is relatively selective to this component.

If $K < 1$, then $y_n < x_n$ and the component is not preferentially passed through the membrane from the high-pressure to the low-pressure side, at least as far as the mole fraction makeup is concerned. However, for the reverse direction, from the low-pressure side to the high-pressure side, the component could be regarded as being preferentially passed through. Whether not this viewpoint is allowable is debatable, of course, since it is contrary to the concept of permeation, where all components move in only one direction, each under a partial pressure or activity difference. This dichotomy, however, sets up the separation in a multistage or cascade operation where internal reflux or recycle occurs.

Substituting for x_n,

$$y_{n+1} = \frac{L}{VK} y_n + \frac{D}{V} x_D$$

$$= A y_n + \frac{D}{V} x_D$$

where $A = L/VK$, as commonly employed in absorber-type calculations, and is called the *absorption factor* (1, 2, 3, 4). Starting at membrane cell $n = 1$, where $y_1 = x_D$,

$$y_2 = \frac{L}{VK} y_1 + \frac{D}{V} y_1 = \left(A + \frac{D}{V} \right) y_1$$

In turn,

$$y_3 = Ay_2 + \frac{D}{V}y_1 = A\left(A + \frac{D}{V}\right)y_1 + \frac{D}{V}y_1$$

$$= \left(A_2 + A\frac{D}{V} + \frac{D}{V}\right)y_1$$

and

$$y_4 = Ay_3 + \frac{D}{V}y_1 = A\left(A^2 + A\frac{D}{V} + \frac{D}{V}\right)y_1 + \frac{D}{V}y_1$$

$$= \left(A^3 + A^2\frac{D}{V} + A\frac{D}{V} + \frac{D}{V}\right)y_1$$

and so on.

It will be observed that, if

$$A + \frac{D}{V} = \frac{L}{VK} + 1 - \frac{L}{V} = 1$$

then $1/K = 1$ or $K = 1$. Under this circumstance, it will be found that $y_1 = y_2 = y_3 = y_4 = \cdots = y_{n+1}$. If, however,

$$A + \frac{D}{V} > 1$$

then $1/K > 1$ or $K < 1$. Under these circumstances, $y_1 < y_2 < y_3 < y_4 < \cdots < y_{n+1}$; that is, the concentration of the component increases going down the rectifying section and decreases going up. This is the circumstance for the less-permeable component j.

If, on the other hand,

$$A + \frac{D}{V} < 1$$

then $1/K < 1$ or $K > 1$. For this opposite circumstance, $y_1 > y_2 > y_3 > y_4 > \cdots > y_{n+1}$. The concentration of the component decreases going down the rectifying section and increases going up. This is the circumstance for the more-permeable component i.

In general, for y_{n+1},

$$y_{n+1} = A^n y_1 + (1 + A + A^2 + A^3 + \cdots + A^{n-1})\frac{D}{V}y_1$$

Furthermore,

$$\frac{1}{1-A} = 1 + A + A^2 + A^3 + \cdots + A^{n-1} + \frac{A^n}{1-A}$$

Therefore,

$$y_{n+1} = A_n y_1 + \frac{(1 - A_n)}{1 - A} \frac{D}{V} y_1$$

$$= \frac{A_n - A_{n+1} + (1 - A_n)\dfrac{D}{V}}{1 - A} y_1$$

Since

$$\frac{D}{V} = \frac{V - L}{V} = 1 - \frac{L}{V}$$

then

$$y_{n+1} = \frac{A^n - A^{n+1} + (1 - A^n)\left(1 - \dfrac{L}{V}\right)}{1 - A} y_1$$

$$= \frac{(1 - A^{n+1}) - (1 - A^n)\dfrac{L}{V}}{1 - A} y_1$$

Accordingly, given n,

$$\sum \frac{y_{n+1}}{\dfrac{(1 - A^{n+1}) - (1 - A^n)\dfrac{L}{V}}{1 - A}} = \sum y_1 = \sum x_D = 1$$

where, at the feed location, for a component i, it can be assumed that $y_{n+1} = Kx_F$; that is, the feed is introduced into the high-pressure or reject side of cell $n + 1$ or cell $m + 1$ and is assumed to be at its "bubble point" and to have the same composition as the reject leaving cell $n + 1$. (Cell $m + 1$ and cell $n + 1$ are the same cell, also called the *feed cell*.)

The solution for this summation is trial and error in L/V, which at the same time establishes the values $(y_1)_i = (x_D)_i$.

Material Balances

Knowing L/V in turn establishes L and D:

$$L = (L/V)(V)$$
$$D = V - L$$

where numerically $V = V'''$, as previously established from a bubble-point type calculation on the feedstream. And, in effect, this value of $V = V'''$ remains the same throughout both the rectifying section and the stripping section. Moreover, this assertion that $V = V'''$ is equivalent to stating that all the membrane cells are to have the same area; that is, if the permeate flow rate and the permeate flux are constant, then the membrane area must remain constant from cell to cell. The analogy is to a distillation column, where the feed is introduced as a saturated liquid of the same composition as the liquid stream leaving the bottom of the feed plate or tray. (As previously noted, this partitioning of the feedstream is a feature of the McCabe-Thiele method for binary distillation calculations.[1])

Constancy

Note that a bubble-point type calculation on the feedstream composition is used to arrive at a value for K_i (or K_j). Albeit this value, in principle, varies from cell to cell as the composition changes, it nevertheless furnishes a means for determining a value. Whereas in vapor-liquid operations such as absorption, the operating temperature and pressure are used to assign a constant value for the liquid-vapor equilibrium vaporization ratio K for a particular component; namely, the key component or components. (And, in general, the equilibrium vaporization ratio is also a function of composition, especially near the critical point of the mixture, and even in absorption, the temperature varies somewhat up and down the column due to enthalpic effects.)

Note also that not only is the feed mixture to be at its permeation bubble point, but it is inferred that the entirety of the reject stream or phase at each cell is at the same bubble-point condition (never mind that the composition—and bubble point—vary). Accordingly, by this simplification, not only does the permeation K-value remain constant, the permeate flux rate V'' has the same constant value from cell to cell, since the other contributing terms P_i, P_L, and P_V remain constant.

The analogy is with equilibrium-stage vapor-liquid operations such as absorption, stripping, or distillation, where the liquid phase is, by definition, considered at its bubble point, that is, at saturation. In distillation, however, there is a marked variation in temperature, which is ordinarily taken into account in the plate-to-plate methods used; that is, in distillation, the equilibrium vaporization ratios are required to vary from plate to plate, at the same time requiring the calculation of a new equilibrium condition at each plate or stage.[1,2]

Feed Dew Point vs. Bubble Point

The feedstream dew-point condition may alternately be calculated. This is equivalent to introducing the feedstream on the permeate side of the membrane cell. In distillation parlance, this is analogous to introducing the feed as a saturated vapor at the feed plate. The McCabe-Thiele method for distillation calculations also accommodates this kind of exigency. For this consequence, $\overline{L} = L$ and $V = F + \overline{V}$.

There may also be a partitioning or proration, whereby the feedstream is subjected to a flash-type calculation, signifying that one part ends up as permeate and the other part as reject. The permeate and reject rates are adjusted accordingly.

Interestingly, according to Example 3.1 of Chapter 3, there is but little variation in the permeate flux V''' when proceeding from the bubble-point type calculation to the dew-point type calculation.

All these aspects tend to signify a great deal of flexibility in adjusting the relative permeate/reject rates in the rectifying and stripping sections; that is, the recycle or reflux ratios can be altered or adjusted internally or externally to favor the degree of separation desired. In other words, these ratios can be regarded as operating parameters and set independent of one another, for both the rectifying and stripping sections. This feature is further emphasized by the ability to control the interstage or intercellular flow rates for both permeate and reject; in fact, they have to be controlled.

Flash-Type Calculations Based on Internal Reflux or Recycle Ratios

As an alternative to using the limiting cases of bubble-point type or dew-point type calculations, the internal recycle or reflux ratio can be used to calculate a value for V''' (and each K_i) via a flash-type calculation.

That is, say, for the rectifying section, the feed to the reject side of a membrane cell can be perceived as the quantity $V + L = F$, or $1 + L/V = F/V$. Since $V = L + D$, then

$$L/V = \frac{1}{1 + \dfrac{1}{L/D}}$$

Therefore, the assigned external reflux ratio L/D can be used to establish L/V.

Accordingly,

$$V/F = \frac{1}{1 + L/V} = \frac{1}{1 + \dfrac{1}{V/L}}$$

Therefore, specifying or knowing L/V or V/L (or L/D) yields V/F, from which a flash-type determination can be made that yields a value for V'''. As already indicated, however, this value expectedly does not vary appreciably using different values of V/F between the bubble-point and the dew-point types of calculation.

Similar relationships may be derived for the stripping section or the rectifying section and can be assumed controlling. In any event, whatever determination is used for establishing V''', it is conceivably not of great consequence, considering all the other assumptions made.

Stage-to-Stage or Cell-to-Cell Calculations

These more rigorous determinations involve a flash-type calculation for each cell, conducted along with material balances. That is, the combined streams $(V + L)$ to each cell can be collectively designated as F, with the flash-type calculation determining the permeate phase compositions and reject phase compositions. The calculation is trial and error for some stream rate or stream ratio in the rectifying section, say, V or L/V, and similarly for these ratios in the stripping section.

To furnish an example, say, for cell number 1 in the rectifying section, here

$$F = L_0 + V_2$$

$$F(x_F)_i = L_0(x_0)_i + V_2(y_2)_i$$

from which a flash-type calculation for cell number 1 can be used to relate the compositions of streams V_1 and L_1 leaving the cell. As developed in Chapter 3, this calculation can be expressed as

$$\sum \frac{(x_F)_i}{\dfrac{L}{F} + \dfrac{V}{F}K_i} = \sum (x_1)_i = 1$$

where

$$K_i = \frac{P_i P_L}{V'' + P_i P_V}$$

and where F may be arbitrarily and temporarily assigned a value of unity for the purposes of only the flash-type calculation. The calculation as stated here can be regarded as trial and error in V'' for an assigned value of V/F, where $L/F = 1 - V/F$. The solution at the same time yields values for the $(x_1)_i$, which can in turn be used to provide values for $(y_1)_i = K_i(x_1)_i$.

Alternately, the calculation can be perceived as determining V'' for an assigned value of V/F or L/F. In other words, either V/F or V'' can be treated as a design-controlled parameter.

In turn, a new value for F for the adjacent stage can be initiated. The logistics of the cell-to-cell calculation sequence are such that it must start either at the top of the rectifying section with an assumed product composition or at the feed location with the reject or permeate phases made equal to the feed composition or else prorated.

It may be emphasized that the streams leaving each cell are in an "equilibrium" condition, so that a dew-point type calculation on stream V_n yields the composition of stream L_n and a bubble-point type calculations on stream L_n yields the composition of stream V_n. This circumstance is built into the flash-type calculation, whereby the two streams leaving are always at "equilibrium" and the compositions are related by K-values.

Partitioning of the Feedstream

As indicated in the previous subsection, the feedstream F can be partitioned or prorated between the reject and permeate sides of the particular cell at the feed location.

(In matter of fact, the feedstream may be prorated into any of the cells, at will, but would involve stage-to-stage or cell-to-cell calculations. However, in distillation, the rule of thumb is that multifeed locations are used only if there are several feedstreams with different compositions. Each feedstream is introduced at the point or location that most likely approximates its own composition. This is a facet of so-called optimum feed location, where the overall separation is made more efficient. The same sort of remarks can be made for intermediate permeate or reject withdrawal, if a stream product of a particular composition is desired.)

The partitioning can be viewed first as follows:

$$\overline{L} = L + XF$$

$$V = \overline{V} + (1 - X)F$$

where X denotes the (molar) fraction of the feedstream F, which is pro-rated to the reject side of the cell, and $(1 - X)$ denotes the fraction prorated to the permeate side.

A material balance around the cell at the feed location can be expressed variously as

$$\overline{L} + V = L + \overline{V} + F$$

$$\overline{L} - \overline{V} = L - V + F$$

$$\left[\left(\frac{\overline{L}}{\overline{V}}\right) - 1\right]\overline{V} = \left[\left(\frac{L}{V}\right) - 1\right]V + F$$

Solving for F,

$$F = \left[\left(\frac{\overline{L}}{\overline{V}}\right) - 1\right]\overline{V} - \left[\left(\frac{L}{V}\right) - 1\right]V$$

Substituting for F into the partition expression involving, say, V and $(1 - X)$ results in

$$\overline{V} = V - (1 - X)\left\{\left[\frac{\overline{L}}{\overline{V}} - 1\right]\overline{V} - \left[\frac{L}{V} - 1\right]V\right\}$$

Collecting terms,

$$\overline{V}\left\{1 + (1 - X)\left[\frac{\overline{L}}{\overline{V}} - 1\right]\right\} = V\left\{1 + (1 - X)\left[\frac{L}{V} - 1\right]\right\}$$

Accordingly,

$$\frac{\overline{V}}{V} = \frac{1 + (1 - X)\left[\frac{L}{V} - 1\right]}{1 + (1 - X)\left[\frac{\overline{L}}{\overline{V}} - 1\right]}$$

It may be readily observed that, when $X = 1$, $\overline{V} = V$ and $\overline{L} = L + F$. And, when $X = 0$, $\overline{V}/V = (L/V)/(\overline{L}/\overline{V}) = (L/\overline{L})(\overline{V}/V)$, whereby $L/\overline{L} = 1$ (and $V = \overline{V} + F$).

The remaining contingency would occur if $L/V = \overline{L}/\overline{V}$. This latter circumstance, however, is not an allowable, unless we consider the limiting condition of total recycle or total reflux.

Proration of Cell Areas

Another facet of assuming a uniform and constant value for K_i throughout is that, other things being equal, the permeate flux also is uniform and constant from cell to cell, in both the rectifying and stripping sections. That is, for a component i, since

$$K_i = \overline{K}_i = \frac{P_i P_L}{V'' + P_i P_V}$$

then $V'' = V/A$ is constant, where A represents the area of each membrane cell. Moreover, since V is constant, A is constant; that is, each cell in the rectifying section has the same transfer area.

With regard to the stripping section, it also follows that $V'' = \overline{V}/\overline{A}$ where $\overline{A}$ is the area of each cell in the stripping section. Therefore,

$$V/A = \overline{V}/\overline{A} \qquad \text{or} \qquad \frac{\overline{A}}{A} = \frac{\overline{V}}{V}$$

Thus, the ratio of the cell areas in the stripping section and rectifying section are the same as the ratio of the respective permeate rates, as previously determined.

Accordingly, establishing the area per cell in the rectifying section permits proration to the area per cell for the stripping section. As the preferable case, if $X = 1$, then $\overline{A} = A$ and the area per cell is the same in both sections, as based upon the permeate.

Proration of Stream Rates

All stream rates may be based on the permeate flux V'' as appears in the K-value denoted K_i in the previous section and elsewhere. Since V'' is to be a constant value common to both the rectifying and stripping sections, other streams may be referenced to this value at steady-state conditions; that is, all stream flow-rate values are also constants.

Hence, we can write that

$$F'' = (F/V)V'' \qquad D'' = (D/V)V'' \qquad L'' = (L/V)V''$$

$$B'' = (B/V)V'' \qquad \overline{V}'' = (\overline{V}/V)V'' \qquad \overline{L} = (\overline{L}/V)V'' = (\overline{L}/\overline{V})(\overline{V}/V)V''$$

Note furthermore that we could speak of the value $\overline{V}'' = \overline{V}/A$ in the terms

$$\overline{V}'' = \overline{V}/A = (V/A)(\overline{V}/V) = \overline{V}''(\overline{V}/V)$$

This is as distinguished from $\overline{V}/\overline{A} = V/A = V''$ used in the preceding subsection. In other words, the latter expression could as well be labeled $\overline{V}'' = \overline{V}/\overline{A} = V/A = V''$ with the overbar or overline extending all the way across, or some other designator could be used. In any event, the calculated permeate flux remains at the same constant value throughout both sections, albeit the molar permeate flow rate can change between the rectifying and stripping sections; that is, if $\overline{V} \neq V$.

As a further note, the feedstream partitioning factor X can be further introduced via the ratio $\overline{V}/V$ (or $\overline{A}/A$). And, for most purposes, however, it is preferable to reference all streams to V'' (or to A) as determined from the K-value determination rather than to $\overline{V}''$ (or to $\overline{A}$).

Stream Flow Rate Consistency

The end products are related by the following balances:

$$V = L + D$$
$$\overline{L} = \overline{V} + B$$

whereby

$$\frac{\overline{L} - B}{L + D} = \frac{\overline{V}}{V} \quad \text{or} \quad \frac{\overline{L}/B - 1}{L/D + 1}\frac{B}{D} = \frac{\overline{V}}{V} = \frac{1 + (1 - X)[L/V - 1]}{1 + (1 - X)[\overline{L}/\overline{V} - 1]}$$

In turn, since $V = L + D$ and $L = V + B$, as shown elsewhere, the external recycle ratios L/D and $\overline{V}/B$ are respectively related to the internal recycle ratios L/V and $\overline{V}/\overline{L}$ by

$$L/V = \frac{1}{1 + \dfrac{1}{L/D}} \quad \text{or} \quad L/D = \frac{1}{\dfrac{1}{L/V} - 1}$$

$$\overline{V}/\overline{L} = \frac{1}{1 + \dfrac{1}{\overline{V}/B}} \quad \text{or} \quad \overline{V}/B = \frac{1}{\dfrac{1}{\overline{V}/\overline{L}} - 1}$$

By virtue of the overall material balances,

$$F = D + B$$
$$F(x_F)_i = D(x_D)_i + B(x_B)_i$$

it follows that

$$\frac{B}{D} = \frac{(x_D)_i - (x_F)_i}{(x_F)_i - (x_B)_i}$$

This relationship is indicative of the following trends, for a given feed composition. For the more-permeable component i, an increase in $(x_D)_i$ is accompanied by an increase in B or a decrease in D or both. Whereas a decrease in $(x_B)_i$ is accompanied by a decrease in B or an increase in D. For the less-permeable component j, the trends are just opposite.

Note, in turn, that

$$\frac{B}{D} = \frac{\overline{L} - \overline{V}}{V - L} = \frac{\overline{L/V} - 1}{1 - L/V} \frac{\overline{V}}{V}$$

$$= \frac{\overline{L/V} - 1}{1 - L/V} \frac{1 + (1 - X)[L/V - 1]}{1 + (1 - X)[\overline{L/V} - 1]}$$

or

$$\frac{(x_D)_i - (x_F)_i}{(x_F)_i - (x_B)_i} = \frac{\overline{L/V} - 1}{1 - L/V} \frac{1 + (1 - X)[L/V - 1]}{1 + (1 - X)[\overline{L/V} - 1]}$$

For a given feedstream composition $(x_F)_i$, given ratios L/V and $\overline{L/V}$ or $\overline{V/L}$, and a partitioning X of the feedstream F, the preceding is an equivalent statement of the overall material balance. And for, say, a particular value of $(x_D)_i$ assumed or calculated, a corresponding value of $(x_B)_i$ satisfies the material balances.

A great advantage, therefore, in using the absorption/stripping factor method, at constant values of the absorption factor A and the stripping factor S, lies in starting the derived stagewise calculation for both the rectifying and stripping sections at the feedstream composition. (Stripping factor derivations are presented in Section 4.5.) Since the derivations for the rectifying and stripping section are but a sequence of material balances, the overall material balance automatically is satisfied, provided each such calculation starts with the feedstream composition or its partitioning.

Effect of Recycle Ratios on Membrane Separations

Note in the foregoing derivations that $L/V < 1$ and $\overline{L/V} > 1$ or $\overline{V/L} < 1$. Furthermore, for the more permeable component i,

As L/V (or L/D) increases, $(x_D)_i$ tends to increase.
As $\overline{L}/\overline{V}$ decreases or $\overline{V}/\overline{L}$ (or $\overline{V}/B$) increases, $(x_B)_i$ tends to decrease.

The less permeable component j acts oppositely.

Effect of X

A further inspection of these trends may be made in terms of X, the (molar) fraction of the feedstream partitioned as reject phase at the feed location: When $X = 1$, it follows that

$$\frac{(x_D)_i - (x_F)_i}{(x_F)_i - (x_B)_i} = \frac{\overline{L}/\overline{V} - 1}{1 - L/V}$$

However, when $X = 0$,

$$\frac{(x_D)_i - (x_F)_i}{(x_F)_i - (x_B)_i} = \frac{\overline{L}/\overline{V} - 1}{1 - L/V} \frac{1 + [L/V - 1]}{1 + [\overline{L}/\overline{V} - 1]} = \frac{\overline{L}/\overline{V} - 1}{1 - L/V} \frac{L/V}{\overline{L}/\overline{V}}$$

Since $L/V < 1$ and $\overline{L}/\overline{V} > 1$, the trend for the more permeable component i is as follows:

As X increases, $(x_D)_i$ increases and $(x_B)_i$ also increases.
As X decreases, $(x_D)_i$ decreases and $(x_B)_i$ decreases.

The converse occurs for the less-permeable component j.

Therefore, it may be concluded that the trends are somewhat offsetting, as far as sharpness of separation is concerned; that is, values of $X \uparrow 1$ favor concentrating the more-permeable component i in the product stream D but at the expense of also raising the concentration in the product stream B. And values of $X \downarrow 0$ favor concentrating the less-permeable component j in the product stream B but at the expense of also increasing the concentration in the product stream D.

A compromise of course is to use $X \sim 0.5$. As an alternative, the number of stages can be increased in either the stripping or rectifying section to counteract whichever trend may occur. That is to say, it may be observed in distillation practice that, if the feedstream is introduced as a liquid (either saturated or supercooled, whereby $X \sim 1$, or $X \geq 1$), there may be a considerably larger stripping section, both in the number of stages or plates and in the column diameter. (The vapor load is also higher in the stripping section if the feedstream is a supercooled liquid: that is, the reboil ratio is correspondingly increased, with the necessary additional heat supplied at the reboiler. In other words, $\overline{V} > V$ if the feedstream is supercooled liquid.) Conversely, if the feedstream is a vapor (saturated or superheated, whereby $X \sim 0$ or $X \leq 0$), the rectifying section is much larger. (This latter circumstance is the more-unusual embodiment. There is a higher cooling load at the reflux condenser, and $L > \overline{L}$.) The same sort of adjustments can be applied to membrane units.

Limiting Values for A

Observe that, for any component i, when $K = 1$ and $A = L/V$, then $y_{n+1} = y_1$; that is, on substituting into the appropriate formula for either component i or component j,

$$y_{n+1} = \frac{\left[1 - \left(\frac{L}{V}\right)^{n+1}\right] - \left[\frac{L}{V} - \left(\frac{L}{V}\right)^{n+1}\right]}{1 - \frac{L}{V}} y_1 = y_1$$

This establishes a crossover for the separation. Below $A = L/V$, the separation goes one way; above $A = L/V$, the separation goes the other way.

Note, furthermore, that when A or $A_i = 1$, the ratio y_{n+1}/y_1 for a component i becomes indeterminate by the previously derived fractional formula. Moreover, it will be found that the derivative dy_{n+1}/dA exhibits infinite behavior.

Interestingly, however, using instead the series form for the solution, when $A = 1$ such that $L/V = K$, it follows that

$$y_{n+1} = y_1 + (n-1)\frac{D}{V}y_1 = \left[1 + (n-1)\frac{D}{V}\right]y_1 \quad \text{or} \quad \frac{y_{n+1}}{y_1} = 1 + (n-1)\frac{D}{V}$$

Since $V = L + D$, then $D/V = 1 - L/V$; and this relationship becomes

$$\frac{y_{n+1}}{y_n} = 1 + (n-1)\left(1 - \frac{L}{V}\right)$$

In another way of viewing the situation, for a particular component i, when A or $A_i = 1$, and introducing the component subscript, then it also follows that

$$(y_{n+1})_i - (y_1)_i = (n-1)(D/V)(y_1)_i$$

For sufficiently large values of n, the difference can be made arbitrarily greater than unity, which is not allowable.

Therefore, for an accumulation of reasons, it can be perceived that there is some sort of a limit on the value or values for A, whether approached from below for component i or from above for component j. For instance, since $A = L/VK$, the larger is the value of K (as for component i), the smaller the value of A. And the smaller the value of K (as for component j), the greater the value of A. These effects are examined further in terms of crossover, where there is a marked change (sea change) between the separation behavior of one component versus the other.

Crossover

An examination of the expression relating $y_{n+1} = Kx_F$ to y_1 establishes certain restrictions for components i and j in the rectifying section, which have to do with the relative magnitude of the mole fractions in the streams entering and leaving the rectifying section. Therefore, it is a separation requirement for component i that

$$y_1 > y_{n+1} \quad \text{or} \quad (y_1)_i > (y_{n+1})_i$$

and for component j that

$$y_1 < y_{n+1} \quad \text{or} \quad (y_1)_i < (y_{n+1})_i$$

This signifies a crossover depending on the magnitude of the absorption factor or A values A_i and A_j and the magnitude of, say, the internal reflux ratio L/V.

That is, consider the previously derived formula in the form

$$y_{n+1} = \frac{A^n - A^{n+1} + (1 - A^n)\left(1 - \dfrac{L}{V}\right)}{1 - A} y_1$$

$$= \frac{(1 - A^{n+1}) - (1 - A^n)\dfrac{L}{V}}{1 - A} y_1$$

It is necessary, therefore, for component i, that

$$\frac{[1 - (A_i)^{n+1}] - [1 - (A_i)^n]\dfrac{L}{V}}{1 - A} < 1$$

and for component j that

$$\frac{[1 - (A_j)^{n+1}] - [1 - (A_j)^n]\dfrac{L}{V}}{1 - A_j} > 1$$

Meeting these requirements requires a judicious choice for the recycle or reflux ratio L/V or L/D, within certain limits.

Note, of course, that magnitude of these expressions is controlled by

$$[1-(A_i)^n]\frac{L}{V} \quad \text{and} \quad [1-(A_j)^n]\frac{L}{V}$$

Speaking only in approximate terms, the inference is that $A_i < {\sim}1$ and $A_j > {\sim}1$. This, however, is only approximately the case, since the internal reflux or recycle ratio L/V has a pronounced influence on the terms

$$A_i = \frac{L}{V}\frac{1}{K_i} \quad \text{and} \quad A_j = \frac{L}{V}\frac{1}{K_j}$$

This effect may be ascertained in the spreadsheet calculations as presented in Appendix 4.

Assignment of a Constant Permeate Rate V

The permeate rate from each cell or stage is assumed to have the same numerical constant value $V = V''$; that is, the system is to be assumed controlled such that the permeate rate is a constant leaving each cell. In other words, referring to Figure 4.2(a), $D = V_1 = V_2 = V_3 = \cdots = V_{n+1} = V$, where V is an assigned constant or parameter.

Furthermore, examining say the second stage, a material balance is

$$V_3 + L_1 = V_2 + L_2$$

or

$$V_3 - V_2 = L_2 - L_1$$

or

$$V_3 - L_2 = V_2 - L_1 = \Delta$$

where Δ must be a constant. This would set up the familiar difference point or "delta" point calculations encountered in multistage separation operations.[2]

Since $V_2 = V_3$, then $L_1 = L_2$. Generalizing, $L_1 = L_2 = L_3 = \cdots = L_{n+1} = L$. There is the qualification, however, that $L_0 = 0$. Also it is understood that $L < V$ in the rectifying section.

However, albeit the permeate flow rates V have the same value in successive stages and the reject flow rates L are the same, the compositions from one stage to another are not the same; that is,

$$(y_1)_i \neq (y_2)_i \neq (y_3)_i \neq \cdots \neq (y_n)_i$$

$$(x_1)_i \neq (x_2)_i \neq (x_3)_i \neq \cdots \neq (x_n)_i$$

Indeed, this is the purpose of multistage separation, to effect composition changes from one stage or cell to another.

Moreover, the convention applies that the quantity $(L_1 + V_2)$ corresponds to a combined feed or feed rate F_2 to the second cell or stage, where L_2 is the reject. (Note that the composition of L_2 and L_1 are different, albeit the stream flow rates L_2 and L_1 have the same value, designated L.) It necessarily follows that $(L_n + V_{n+1}) = F_n$ is constant from stage to stage, since the values of L remain the same constant value and the values of V are the same constant value. In other words, here, ignoring the first stage, $F_2 = F_3 = \cdots = F_n$. The (combined) compositions, however, vary.

From the derivations and calculations for a single cell or stage as developed in Chapter 3, the molar permeate rate V (as well as the reject rate L and combined feed rate F) properly must vary from stage-to-stage, since the (combined) feed composition also varies. This implies that something has to give. For example, if the permeate rate is assumed constant, then the mole fraction summations cannot be held exactly to unity. Or, if the mole fraction summations are required to be unity, then the permeate rate should be allowed to vary. The former course is pursued here in the further interests of simplification.

From another standpoint, if the rigorous flash-type calculation, say, is to be used at each stage for a fixed vapor to combined feed ratio, then V'' has to vary as the composition of the combined feed input to the cell varies. In turn, the K-values would vary.

Interestingly, as shown in Example 3.1 at least, the calculated permeate flux V'' does not vary appreciably, even as the parameter V/F is changed. Nor do the resulting permeate and reject compositions change appreciably. That is, at least in this particular example, there are only minimal changes in composition.

It may be added that the V/F ratio, as developed in Chapter 3, relates to the reflux ratio L/V or L/D, as described herein in Chapter 4. That is, V/F, as per Chapter 3, corresponds to $L/(V + L)$, as per Chapter 4. That is to say,

$$\frac{V}{F} = \frac{L}{V + L} \quad \text{or} \quad V/F = \frac{1}{1 + \dfrac{1}{L/V}}$$

where the derivation in Chapter 3 utilizes V/F as a parameter. There, stream F consists of a single stream introduced into the reject side of the membrane. Calculating V by the trial-and-error procedure utilized then ultimately permits the determination of the corresponding value for F (and L).

On the other hand, when applied to multistage operations, the concept of F pertains to the introduction of both V and L (that is, to $V + L$) to the reject side of the membrane, where V and L originate from the adjacent membrane cells, that is, from the posterior and anterior cells. Thus, the so-called internal reflux ratio or recycle ratio L/V can be used to establish a value for V/F, from which a value for the permeate flux V'' could in turn be calculated by the same methods of Chapter 3. The value of V'' so determined is the uniform and constant permeate rate at each stage in the rectifying section. This would supersede the determination of the permeate flux V'' based on the feedstream per se.

The relationship between the internal reflux ratio and the external reflux ratio L/D is given starting with $V - L = D$, where

$$L/D = \frac{1}{\dfrac{1}{L/V}} - 1 \quad \text{or} \quad L/V = \frac{1}{\dfrac{1}{L/D} + 1}$$

This is the familiar reflux relationship as applies notably to distillation.

Relaxation of the Mole Fraction Summation Requirement

As a final note, the relaxation of the requirement that the mole fractions of a stream sum to unity is inherent in the utilization of the absorption-stripping factor concept as developed here. Ordinarily, the concept is used in practice where an absorbed or stripped component is in relatively low concentrations in, say, the absorbing phase. This is most evident in the use of an absorber oil or lean oil (L.O.) to absorb lower molecular weight hydrocarbons from a natural gas stream; that is, there occurs in ascending order the absorption of some methane, more of ethane, most of the propane and butanes, and substantially all of the pentanes, hexanes, and heavier hydrocarbons. Other things being equal, the lower the vaporization ratio, volatility, or K-value of the component, the higher the value for the absorption factor $A = L/VK$ and the greater the degree of absorption that takes place. This can be seen from the absorption factor plot contained in the standard references,[1,2,5,6] as originally derived by Souders and Brown.[7] This trend can also be ascertained from the previous derivations in comparing y_{n+1} with y_1 (where the greater the value for A, the greater the difference between y_{n+1} and y_1 or the greater the ratio y_{n+1}/y_1) or by still other arrangements or rearrangements.

The bottoms product mixture from the absorber is characterized as rich oil, and the absorbed hydrocarbons are steam-stripped, leaving the lean oil phase for recycle to the absorber (the lean oil is generally characterized as an averaged 180 molecular weight paraffinic-type oil). The light hydrocarbons so stripped are, in turn, concentrated in successive distillation columns variously called a *deethanizer, depropanizer, deisobutanizer,* or *debutanizer,* with the pentanes and heavier hydrocarbons largely making up a natural gasoline fraction. The operating pressure of each of these columns is dictated by the necessity of condensing a reflux phase, usually by the use of cooling water in a condensing heat exchanger.

It may be added that the sharp separation of methane and ethane from the other absorbed hydrocarbon components ordinarily requires low-temperature or cryogenic distillation operations to produce a reflux phase at the top of the column. Alternately, extractive distillation methods can be used, where an absorber oil is introduced at the top of the column instead of reflux. The degree of separation can be enhanced if the reboiler at the bottom of the column is replaced by a separate distillation-type column, with the overhead from this column recycled to the bottom of the first column, an operation once called the *extractifrac process.*[1]

The absorber gaseous product, or off-gas, composed mostly of methane with some ethane and relatively minor amounts of propane and heavier hydrocarbons, can be characterized as lean natural gas, with the composition adjusted at the absorber so as to meet Btu requirements, if any. The gas is generally dried to meet pipeline specifications, either by glycol treatment or solid adsorbents. If the acid gas hydrogen sulfide is present, removal by absorption with a solution of one or another of the ethanolamines is standard operating procedure. Appreciable concentrations of the acid gas carbon dioxide may be removed similarly. Nitrogen, if at sufficiently low levels, is ignored—if it does not adversely affect the Btu rating; otherwise, the gas is not marketable. All of which gets around to the use of membrane technology to upgrade subquality natural gas by (partially) removing the nitrogen content.

4.5 STRIPPING SECTION

For the stripping section, for a component i, and dropping the subscript,

$$\overline{L}_{m+1}\overline{x}_{m+1} - \overline{V}_m\overline{y}_m = Bx_B$$

where

$$\bar{L}_{m+1} - \bar{V}_m = B$$

If the $\bar{L}_m$ and $\bar{V}_m$ are constants, independent of the cell number, then $K_m = \bar{K}_m$ is a constant, and

$$\bar{L}\bar{x}_{m+1} - \bar{V}\bar{y}_m = Bx_B$$

or

$$\bar{x}_{m+1} = \frac{\bar{V}}{\bar{L}}\bar{y}_m + \frac{B}{\bar{L}}x_B$$

This is a step function which can be represented as a straight line in y-x space. Moreover,

$$\bar{L}/\bar{V} > 1$$

The membrane rate balance for a component i is

$$\bar{V}\bar{y}_m = P_i(\bar{P}_L\,\bar{x}_m - \bar{P}_V\,\bar{y}_m)$$

or

$$\bar{y}_m = \frac{P_i\bar{P}_L}{\bar{V} + P_i\bar{P}_V}\bar{x}_m = \bar{K}\bar{x}_m$$

This also represents a step function, which can be represented as a straight line in $\bar{y}-\bar{x}$ space, with slope $\bar{K}$ and intercept at the origin.

If $\bar{V} = V$, and if $\bar{P}_L = P_L$ and $\bar{P}_V = P_V$, then $\bar{K} = K$.

Substituting for $\bar{y}_m$, where $x_B = \bar{x}_1$, gives

$$\bar{x}_{m+1} = \frac{\bar{V}\bar{K}}{\bar{L}}\bar{x}_m + \frac{B}{\bar{L}}\bar{x}_1$$

$$= S\bar{x}_m + \frac{B}{\bar{L}}\bar{x}_1$$

where $S = \bar{V}\bar{K}/\bar{L}$, as defined by the substitution. The entity S corresponds to the stripping factor in vapor-liquid absorber/stripper calculations.

Therefore,

$$\bar{x}_2 = S\bar{x}_1 + \frac{B}{\bar{L}}\bar{x}_1 = \left(S + \frac{B}{\bar{L}}\right)\bar{x}_1$$

$$\bar{x}_3 = S\bar{x}_2 + \frac{B}{\bar{L}}\bar{x}_1 = S\left(S + \frac{B}{\bar{L}}\right)\bar{x}_1 + \frac{B}{\bar{L}}\bar{x}_1$$

and

$$\overline{x}_4 = S\overline{x}_3 + \frac{B}{L}\overline{x}_1 = S\left(S^2 + S\frac{B}{L} + \frac{B}{L}\right)\overline{x}_1 + \frac{B}{L}\overline{x}_1$$

$$= \left(S^3 + S^2\frac{B}{L} + S\frac{B}{L}\right)\overline{x}_1$$

etc.

Similarly to the derivations for the rectifying section, it will be found that for the stripping section, if

$$S + \frac{B}{L} = \frac{\overline{V}\,\overline{K}}{\overline{L}} + 1 - \frac{\overline{V}}{\overline{L}} = 1$$

then $\overline{K} = 1$ and

$$\overline{x}_1 = \overline{x}_2 = \overline{x}_3 = \overline{x}_4 = \cdots = \overline{x}_{m+1}$$

If

$$S + \frac{B}{L} > 1$$

then $\overline{K} > 1$ and

$$\overline{x}_1 < \overline{x}_2 < \overline{x}_3 < \overline{x}_4 < \cdots < \overline{x}_{m+1}$$

the composition of the component will increase going up the stripping section, and decrease going down.

If

$$S + \frac{B}{L} < 1$$

then $\overline{K} < 1$ and

$$\overline{x}_1 > \overline{x}_2 > \overline{x}_3 > \overline{x}_4 > \cdots > \overline{x}_{m+1}$$

the composition of the component will decrease going up the stripping section, and increase going down.

In general it will be found, as was done for the rectifying section, that

$$\overline{x}_{m+1} = \frac{(1 - S^{m+1}) - (1 - S^m)\dfrac{\overline{V}}{\overline{L}}}{1 - S}\overline{x}_1$$

Accordingly,

$$\sum \frac{\overline{x}_{m+1}}{\dfrac{(1-S^{m+1})-(1-S^m)\dfrac{\overline{V}}{\overline{L}}}{1-S}} = \sum \overline{x}_i = \sum x_B = 1$$

where, at the feed location, $(\overline{x}_{m+1})_i = (y_{n+1})_i/\overline{K}_i$ will be known from the calculations for the rectifying section. Moreover, it may be assumed that $(x_F)_i = (\overline{x}_{m+1})_i$, or at least the approximation can be made.

Stripping Section Ratios vs. Rectifying Section Ratios vs. Feed Introduction

As previously indicated, as per Figure 4.3, for the circumstance where the feed stream F is introduced into the reject side of cell $n + 1$ (or cell $m + 1$, or cell f) and combined with stream $L_{n+1} = \overline{L}_{m+1}$, it will follow that $\overline{L} = L + F$, $\overline{V} = V$. By analogy with distillation, this would correspond to the feed mixture being at its bubble-point type condition.

Alternately, the feed mixture F can be considered introduced into the permeate side of membrane cell $n + 1$ (or $m + 1$) and combined with V_{n+1}, whereby $\overline{L} = L$ and $\overline{V} = V + F$. This would correspond to the feed mixture being at its dew-point type condition.

Or some combination of the foregoing could be utilized, as per the McCabe-Thiele method for binary distillation calculations.[1,2] In membrane units, moreover, the overall operation is subject to the arbitrary regulation of interstage flow rates (and compression rates), according to the discretion of the operator. In any event, the inference is that the internal recycle ratio for the stripping section can be affixed independently of the internal recycle or reflux ratio for the rectifying section. This is not necessarily the case for distillation, however, where the one tends to be dependent upon the other, at least for calculation purposes. Practically speaking, in distillation the behavior of the rectifying section is in part governed by the heat removed at the overhead condenser (inducing the external reflux ratio), and the behavior of the stripping section is in part governed by the heat added at the reboiler (inducing the external reboil ratio), albeit the phenomena are also entwined. The behavior at the feed location, for a given feedstream condition, will then function dependently.

Lastly, as has been previously indicated, it will be assumed that the permeate flux V''' remains at the same constant value throughout both the rectifying and stripping sections. In turn, $\overline{K} = K$.

Furthermore, for calculational purposes, the permeate flux V'' may be numerically equated with V, and in turn the other stream sizes or flow rates related to V''.

Overall Balance Requirements

From the stripping section material balances,

$$\overline{L} = [1/(\overline{V}/\overline{L})]\overline{V} \qquad \text{and} \qquad B = \overline{L} - \overline{V}$$

where it is assumed that $\overline{V} = V''$, and the ratio $(\overline{V}/\overline{L})$ is known or specified. In this way, a corresponding value for B can be calculated.

Finally, since D has been determined for the rectifying section, a value for F will follow:

$$F = D + B$$

and as a check, for each component i, dropping subscripts,

$$Fx_F = Dx_D + Bx_B$$

where the sum of the mole fractions should be unity, for each stream.

Significantly, the assumption that $\overline{V} = V''$ implicitly states that each membrane cell will have the same area. That is, if the permeate rate is constant, and the permeate flux is constant, then the cell area remains constant from cell-to-cell.

Limiting Values for S

By the procedures used to determine the qualifications for A, limiting values for S can be perceived in terms of $S = \overline{V}K/\overline{L}$. Whenever $\overline{K} = K = 1$, then it follows that $S = \overline{V}/\overline{L}$, and substitution into the appropriate form, previously derived, gives

$$\overline{x}_{m+1} = \frac{\left[1 - \left(\dfrac{\overline{V}}{\overline{L}}\right)^{m+1}\right] - \left[\dfrac{\overline{V}}{\overline{L}} - \left(\dfrac{\overline{V}}{\overline{L}}\right)^{m+1}\right]}{1 - \dfrac{\overline{V}}{\overline{L}}} \overline{x}_1 = \overline{x}_1$$

This establishes the circumstance where no separation would occur. For values of S greater than $\overline{V}/\overline{L}$, the separation goes one way, for values less the separation goes the other way.

Crossover

The requirements for separation in the stripping section can be approached in the same manner as for the rectifying section. Consider, therefore, the previous equation, that for no separation to occur,

$$
\overline{x}_{m+1} = \frac{\left[1-\left(\dfrac{\overline{V}}{\overline{L}}\right)^{m+1}\right]-\left[\dfrac{\overline{V}}{\overline{L}}-\left(\dfrac{\overline{V}}{\overline{L}}\right)^{m+1}\right]}{1-\dfrac{\overline{V}}{\overline{L}}}\,\overline{x}_1 = \overline{x}_1
$$

Thus it is a separation requirement for component i that

$$
\overline{x}_1 < \overline{x}_{m+1} \qquad \text{or} \qquad (\overline{x}_1)_i < (\overline{x}_{m+1})_i
$$

and for component j that

$$
\overline{x}_1 > \overline{x}_{m+1} \qquad \text{or} \qquad (\overline{x}_1)_j > (\overline{x}_{m+1})_j
$$

The foregoing will signify a crossover depending upon the magnitude of the stripping factor or S values S_i and S_j, that is, upon the magnitude of the internal reflux ratio $\overline{V}/\overline{L}$ and/or the K-values.

Therefore, based on the previously derived formula, that

$$
\overline{x}_{m+1} = \frac{(1-S^{m+1})-(1-S^m)\dfrac{\overline{V}}{\overline{L}}}{1-S}\,\overline{x}_1
$$

it is necessary for component i that

$$
\frac{[1-(S_i)^{m+1}]-[1-(S_i)^m]\dfrac{\overline{V}}{\overline{L}}}{1-S_i} > 1
$$

and for component j that

$$
\frac{[1-(S_j)^{m+1}]-[1-(S_j)^m]\dfrac{\overline{V}}{\overline{L}}}{1-S_j} < 1
$$

Meeting these requirements would ordinarily require a judicious choice for the stripping section recycle or reboil ratio expressed as $\overline{V}/\overline{L}$ or $\overline{V}/B$, within certain limits. This in turn would have to be interfaced with the results for the rectifying section, introducing an additional element or elements of trial and error.

Fortunately, however, these stripping section ratios or flow rates can be made dependent upon the rates in the rectifying section, as follows.

Assumption of a Constant Permeate Rate

As developed in the preceding section for rectification, if the permeate molar flow rate $\overline{V} = V$ is assumed to have the same constant value throughout the stripping section or is held constant, so are the reject rates and the combined permeate and reject rates (or feed rates) to each cell or stage; moreover, $\overline{L} > \overline{V}$.

Furthermore, if this constancy is assumed, then something else must be allowed to give. A rigorous calculation for each stage then requires that the permeate fluxes $\overline{V}''$ and V'' vary from stage to stage and, in turn, the K-values, along with composition. The alternative is to assume that the mole fractions do not necessarily sum to unity. This latter course is henceforth pursued as the simpler route, where the permeate flux and the K-values are to be assumed constant and uniform throughout each section.

More Rigorous Statement

For the more rigorous interpretation, it can be emphasized that, for the stripping section, the V/F or $\overline{V}/F$ ratio as developed in Chapter 3 corresponds to $\overline{V}/(\overline{V} + \overline{L})$ as described here, in Chapter 4. That is, V/F or $\overline{V}/F$ as per Chapter 3 corresponds to $\overline{V}/(\overline{V} + \overline{L})$ as used herein, in Chapter 4. In other words, $(\overline{V} + \overline{L})$ is the total or combined feed input to each stage. (Note also that the reject from the same stage is also designated $\overline{L}$, albeit it necessarily has a different composition but the same rate. More properly, the terms should be subscripted by stage number.)

In another way of looking at it, under this more general or rigorous interpretation, the total or combined feed to each stage is regarded as "flashed" at the specified vapor to feed ratio. Moreover, the mole fractions of all the streams then are required to sum to unity, as presented in Chapter 3. The multistage calculation, however, has to proceed from cell to cell or stage to stage.

To continue, for the stage-to-stage flash-type calculation,

$$\frac{\overline{V}}{F} = \frac{\overline{V}}{\overline{L} + \overline{V}} \quad \text{or} \quad \overline{V}/F = \frac{1}{\dfrac{1}{\overline{V}/\overline{L}} + 1}$$

where the derivation in Chapter 3 applies on the left-hand side of each form of the equation and the derivation here applies on the right-hand side. Thus, the so-called internal reflux ratio or recycle ratio $\overline{L}/\overline{V}$ or

$\overline{V}/\overline{L}$ (where $\overline{L} > \overline{V}$) can be used to establish a value for $\overline{V}/F$, from which a value for $\overline{V}$ can, in turn, be calculated by the methods of Chapter 3. The value of $\overline{V}$ so determined is the uniform and constant permeate rate for each stage in the stripping section.

Internal vs. External Recycle

As to the relationship relating internal recycle or reflux and external reflux or as pertains to the stripping section, it is based on the material balance statement $\overline{L} - \overline{V} = B$, where

$$\overline{V}/B = \frac{1}{\dfrac{1}{\overline{V}/\overline{L}} - 1} \qquad \text{or} \qquad \overline{V}/\overline{L} = \frac{1}{\dfrac{1}{\overline{V}/B} + 1}$$

This is the relationship between internal and external reboil ratios as they pertain to distillation, which also applies here to the stripping section in multistage membrane separations.

Stripping Section Flow Rates and Ratios

The molar stream flow rates in the stripping section can be made dependent on the rectifying section flow rates. This dependency is made by assuming that the relationship between the product streams D and B can be based on the feedstream composition, in that, ideally, all of component i makes up stream D and all of component j makes up stream B. Admittedly, this is a simplification, but the alternative is an increasing complexity that is probably not justified.

The initial starting point rests on the proposition that an absolute value for V, the molar vapor rate, can be established by a single-stage flash-type calculation on the feed or feedstream to the operation. As developed in Chapter 3 and Appendix 3, this determination is trial and error in V; that is, V is the permeate phase arising from the feed stage designated both by $n + 1$ and $m + 1$ (or by f).

The feedstream to the overall operation is generally designated F but more properly can be distinguished from the feed F to the single-stage separation. The distinguishing criterion is in the context of usage; that is, whether F designates the feedstream for the multistage operation or the combined feed into a single stage (which may be a stage in a multistage of operation). The overall material balance for the multistage operation is

$$\text{Feed} = D + B \qquad \text{or} \qquad F = D + B$$

where the latter is the more conventional notation.

Furthermore, by definition, the feedstream combines with and has the same composition as the reject phase from the feed stage. Also, the permeate phase flow rate V has the same constant value for both stripping and rectification; that is,

$$\overline{V} = V \qquad \text{whereby} \qquad \overline{L} = L + \text{Feed}$$

All the feedstream is introduced into the reject side of the feed stage membrane cell.

Starting, therefore, with an assigned value for the external reflux ratio L/D for the rectifying section, and since $V = L + D$, it follows that

$$L/V = \cfrac{1}{\cfrac{1}{L/D + 1}}$$

As demonstrated elsewhere, it is understood that each L and V remain at the same constant value throughout the rectifying section. In turn,

$$\frac{V}{F} = \frac{V}{V + L} = \cfrac{1}{1 + \cfrac{L}{V}}$$

where this value of V/F is the value used for the trial-and-error, single-stage flash calculation as set forth in Chapter 3 and Appendix 3; that is, $V + L$ represents the combined input into the reject side of the feed stage as diagrammed in Figure 4.3. (The output from the reject side also is designated L, even though it has a different composition than the input entering from the previous stage. The flow rates are merely assumed the same.) This use of F is distinguished from the feedstream F into the overall operation, as previously noted.

The single-stage flash-type calculation will establish an absolute value for V in consistent units. Knowing V, then L can be calculated from the internal reflux ratio L/V, and D can, in turn, be calculated from the initially assigned value for the external reflux ratio L/D.

If it is presupposed or assumed that all the more-permeable component i ends up as the product stream D and all the less-permeable component j ends up as the product B, then

$$B \sim D \frac{(x_F)_j}{(x_F)_i}$$

In consequence, since $\overline{V} = V$, the external reboil or recyle ratio $\overline{V}/B$ can be determined. Furthermore, since $\overline{L} = \overline{V} + B$, then $\overline{V}/\overline{L}$ can be determined directly or calculated from the expression

$$\frac{\overline{V}}{\overline{L}} = \frac{1}{\dfrac{1}{\overline{V}/B} + 1}$$

Thus, stream values and stream ratios can be determined for both the rectifying and stripping sections, starting from an assigned or trial value for L/D. These procedures are applied in Example 4.1.

However, a more rigorous procedure is presented in the following subsection and utilized in the spreadsheet calculations of Appendix 4.

Closure of the Material Balance: The Relationship between L/V and $\overline{L}/\overline{V}$

More rigorously speaking, if the feedstream composition is to be made identical to the reject composition for stage $n + 1 = m + 1$, then there is a relationship between $\overline{L}/\overline{V}$ and L/V that involves the composition of product streams B and D. That is to say, if L/V and x_D are determined by a specification and calculation on the rectifying section for an integral number of stages, then this dictates a constraint between $\overline{L}/\overline{V}$ and x_B. At the same time, the overall material balance is automatically satisfied.

This can be visualized graphically in Figure 4.7 in terms of component i, where the positioning of the operating line for the rectifying section results in an integral number of stages, culminating at point D or x_D. Note, however, that the operating line can be positioned up or down such that the slope remains the same but the intersections at either end will change. The net result is that an integral number of stages can occur between the feed location and point D. While the determination appears to be trial and error in the graphical representation, it can be performed analytically (and exactly) by the absorption factor method, as previously presented.

In turn, the operating line for the stripping section intersects the operating line for the rectifying section at $x = x_F$. Moreover, this operating line culminates at point B or x_B in an integral number of stages. And, as can be noted from Figure 4.7, the terminus at the feed location is fixed, with the necessity of varying the slope to arrive at point B or x_B in the integral number of stages. Although this determination can be performed analytically for an integral number of stages by the stripping factor method, as previously presented, it is not known in advance what value of $\overline{L}/\overline{V}$ to use to be consistent with the value of x_B attained or vice versa. Thus, the element of trial and error is introduced.

The necessary material-balance relationships are derived as follows. For the rectifying section, dropping component subscripts,

$$V = L + D$$

$$Vy = Lx + Dx_D$$

where

$$y = \frac{L}{V}(x - x_D) + x_D$$

For the stripping section,

$$\overline{L} = \overline{V} + B$$

$$\overline{L}\overline{x} = \overline{V}\overline{y} + Bx_B$$

where

$$\overline{y} = \frac{\overline{L}}{\overline{V}}(\overline{x} - x_B) + x_B$$

Since $\overline{y} = y$ and $\overline{x} = x = x_F$ at the point of intersection, then it follows that

$$\frac{\overline{L}}{\overline{V}} = \frac{L}{V}\frac{x_F - x_D}{x_F - x_B} + \frac{x_D - x_B}{x_F - x_B}$$

From having already determined L/V and x_D, this establishes a relationship between $\overline{L}/\overline{V}$ and x_B. Solving for x_B, an alternative arrangement is

$$x_B = \frac{(\overline{L}/\overline{V} - L/V)x_F + (L/V - 1)x_D}{\overline{L}/\overline{V} - 1}$$

where $\overline{L}/\overline{V} > 1$ and $L/V < 1$. On assuming a value for $\overline{L}/\overline{V}$ (or $\overline{V}/\overline{L}$ or $\overline{V}/B$) and calculating a corresponding value for x_B for an integral number of stages via the stripping factor relationship, the value of x_B so calculated must agree with the preceding value of x_B, at least with some margin of error.

Interestingly, this relationship satisfies the overall material balance. Multiplying both the numerator and denominator by $V = \overline{V}$ gives

$$x_B = \frac{(\overline{L} - L)x_F + (L - V)x_D}{\overline{L} - \overline{V}}$$

where $\overline{L} - \overline{V} = B$ and $(L - V) = -D$ and $\overline{L} - L = F$. Therefore,

$$Fx_F = Dx_D + Bx_B$$

which automatically holds true regardless of the values of $\overline{L}/\overline{V}$ vs. x_B used for the stripping section.

4.6 STRIPPING SECTION VS. RECTIFYING SECTION

Note that either the stripping section or the rectifying section may be operated alone. For the former, the feed F becomes stream $\overline{L}_{m+1}$. The bottom product $\overline{L}_1 = B$ is distinctly rich in component j. The other product stream is $\overline{V}_m$, which shows a "sloppy" separation between i and j.

For operation as a rectifying section alone, the feedstream becomes V_{n+1}. The top product $V_1 = D$ is rich in component i or I. The other product L_n shows a sloppy separation between the components.

For a sharp separation between both (key) components, both stripping and rectifying sections must be linked together.

4.7 FEED LOCATION

If the rectifying and stripping sections are linked together, then at the feed membrane cell, for a component i,

$$Vy_{n+1} = P_i(\overline{P}_L \overline{x}_{m+1} - P_V y_{n+1})$$

or

$$y_{n+1} = \frac{P_i \overline{P}_L}{V + P_i P_V} \overline{x}_{m+1} = \overline{K} \overline{x}_{m+1} = K \overline{x}_{m+1}$$

where membrane cell $n + 1$ is the same as membrane cell $m + 1$ and can be called the *feed membrane cell f*.

For the purposes here, the feed composition can be made equal to that of either $\overline{L}_{m+1}$ or V_{n+1}, or the other way round, and combined with either stream. For the first option, which is preferred,

$$L = F + \overline{L} \qquad \text{and} \qquad \overline{V} = V$$

For the second option,

$$V = F + \overline{V} \quad \text{and} \quad \overline{L} = L$$

The situation is analogous to that of a distillation column assuming constant molal overflow.

Alternately, the feedstream could be partitioned between V and $\overline{L}$. Note therefore that, if

$$y_{n+1} = \overline{K}\,\overline{x}_{m+1} = \overline{K}x_F$$

then

$$1 = \sum y_{n+1} = \sum \overline{K}(x_F)_i$$

where

$$\overline{K}_i = K_i = \frac{P_i \overline{P}_V}{\overline{V} + P_i \overline{P}_V} = \frac{P_i P_V}{V + P_i P_V}$$

This expression furnishes the solution for $\overline{V} = V$. Furthermore, for this particular circumstance, $\overline{K} = K$. (Note that $V \equiv V''$.)

For this very reason, it is more convenient to make the composition of stream F identify with that of stream $\overline{L}_{m+1}$ or vice versa and to combine stream F with stream $\overline{L}_{m+1}$ at the cell or leaving the cell. The exact configuration is academic, in theory at least, since perfect mixing is assumed.

In effect, the preceding is the "bubble-point" calculation for the feed. Moreover, the values for $(y_{n+1})_i$ are obtained simultaneously during the process of calculation.

4.8 SEPARATION REQUIREMENTS

It is required that

$$A_i < 1 \quad A_j > 1$$

$$S_i > 1 \quad S_j < 1$$

Furthermore, from the bubble-point type calculation for the feed,

$$K_i > 1 \quad K_j < 1$$

where i is the more "volatile" component, that is, has the higher permeability.

These requirements, in turn, place certain restrictions on the behavior of the permeability and pressure drop with respect to L or vice versa:

$$P_i P_L > V + P_i P_V \quad \text{or} \quad V < P_i (P_L - P_V)$$

$$P_i \overline{P}_L > \overline{V} + P_i \overline{P}_V \quad \text{or} \quad \overline{V} < P_i (\overline{P}_L - \overline{P}_V)$$

$$P_i P_L < V + P_i P_V \quad \text{or} \quad V > P_i (P_L - P_V)$$

$$P_i \overline{P}_L < \overline{V} + P_i \overline{P}_V \quad \text{or} \quad \overline{V} > P_i (\overline{P}_L - \overline{P}_V)$$

where usually $P_V \sim \overline{P}_V$ and $P_L \sim \overline{P}_L$, and for the case at hand, let $\overline{V} = V$.

4.9 TOTAL REFLUX

At a condition of total reflux, $L/V = 1$ and $\overline{V}/\overline{L} = 1$. The following simplifications occur.

Rectifying Section

The summation reduces as follows, where $K_i(x_F)_i = (y_{n+1})_i$. Dropping the component subscript,

$$\sum \frac{K(x_F)(1 - 1/K)}{-(1/K)^{n+1} + (1/K)^n} = \sum y_i = \sum (x_D) = 1$$

Given the K_i, it takes trial and error to determine n, the number of cells or stages in the rectifying section.

Stripping Section

The summation reduces as follows, where $(x_F)_i = (\overline{x}_{m+1})_i$:

$$\sum \frac{(x_F)(1 - K)}{-(K)^{m+1} + K^m} = \sum \overline{x}_i = \sum (x_B) = 1$$

Given the value of K_i, it takes trial and error to determine m, the number of cells or stages in the stripping section.

Total Number of Cells

Counting the feed cell, the total number of cells or stages is $n + m + 1$.

Overall Component Balances

The solutions at total reflux also establish the compositions $(x_D)_i$ and $(x_B)_i$. Accordingly, by the overall component material balances, it is required that, for any two components i and j,

$$\frac{B}{D} = \frac{(x_D)_i - (x_F)_i}{(x_F)_i - (x_B)_i} = \frac{(x_D)_j - (x_F)_j}{(x_F)_j - (x_B)_j}$$

If this condition is not met, then new values for K_i must be chosen, which in turn yield new values for n and m and so forth. The overall solution takes double trial and error. Note that, for more than two components, a complete and rigorous solution may not exist; that is, there will be too many constraints, as evidenced by the overall component material balances. The solution, therefore, should be confined to only the two key components.

4.10 MINIMUM REFLUX

At minimum reflux, it is required that $n \to \infty$ and $m \to \infty$.

Rectifying Section

If $A < 1$, then the summation becomes

$$\sum \frac{K(x_F)\left(1 - \dfrac{L}{VK}\right)}{1 - \dfrac{L}{V}} = \sum \overline{x}_1 = \sum (x_B) = 1$$

Given the value of K_i, it takes trial and error to determine L/V.

Stripping Section

If $S < 1$, then the summation becomes

$$\sum \frac{(x_F)\left(1 - \dfrac{\overline{V}K}{\overline{L}}\right)}{1 - \dfrac{\overline{V}}{\overline{L}}} = \sum \overline{x}_1 = \sum (x_B) = 1$$

Given the value of K_i, it takes trial and error to determine $\overline{V}/\overline{L}$.

Overall Component Balances

The solutions for minimum reflux also establish the values of $(x_D)_i$ and $(x_B)_i$. As before, it is required that for any two components i and j,

$$\frac{B}{D} = \frac{(x_D)_i - (x_F)_i}{(x_F)_i - (x_B)_i} = \frac{(x_D)_j - (x_F)_j}{(x_F)_j - (x_B)_j}$$

If this condition is not met, assume new values for K_i. The overall solution then takes double trial and error. For the reasons stated previously, such as for total reflux, the solution should be confined to the two key components.

4.11 SIMPLIFICATIONS

A simplification can be made for two components that correspond to the condition of minimum reflux. The value of K for each component is assumed uniform throughout.

For sufficiently large values of n and sufficiently small values of $A = L/VK$,

$$(x_D)_i \sim \frac{K_i(x_F)_i}{1 - \dfrac{L}{V}}\left(1 - \frac{L}{VK_i}\right)$$

$$(x_D)_j \sim \frac{K_j(x_F)_j}{1 - \dfrac{L}{V}}\left(1 - \frac{L}{VK_j}\right)$$

where

$$\frac{(x_D)_i}{(x_D)_j} = \frac{\dfrac{K_i V}{L} - 1}{\dfrac{K_j V}{L} - 1}\frac{(x_F)_i}{(x_F)_j}$$

Similarly, for sufficiently large values of m and sufficiently small values of $S = VK/L$,

$$(x_B)_i = \frac{(x_F)_i}{1 - \dfrac{V}{L}}\left(1 - \frac{VK_i}{L}\right)$$

$$(x_B)_j = \frac{(x_F)_j}{1 - \dfrac{V}{L}}\left(1 - \frac{VK_j}{L}\right)$$

where

$$\frac{(x_B)_i}{(x_B)_j} = \frac{1 - \dfrac{\overline{V}K_i}{\overline{L}}}{1 - \dfrac{\overline{V}K_j}{\overline{L}}} \frac{(x_B)_i}{(x_B)_j}$$

Furthermore,

$$\frac{(x_D)_i}{(x_D)_j} = \frac{\dfrac{K_iV}{L} - 1}{\dfrac{K_jV}{L} - 1} \frac{1 - \dfrac{K_j\overline{V}}{\overline{L}}}{1 - \dfrac{K_i\overline{V}}{\overline{L}}} \frac{(x_B)_i}{(x_B)_j}$$

This expression relates the separations that can be attained.
Alternately,

$$\frac{(x_D)_i}{(x_D)_j} = \frac{\dfrac{1}{A_i} - 1}{\dfrac{1}{A_j} - 1} \frac{1 - S_i}{1 - S_j} \frac{(x_B)_i}{(x_B)_j}$$

$$= M\frac{(x_B)_i}{(x_B)_j}$$

where M is defined by the substitution. Since, for a binary system, the two mole fractions for each stream sum to unity, $(x_B)_j$, say, can be solved in terms of $(x_D)_i$:

$$(x_B)_i = \frac{(x_D)_i}{(x_D)_i + M[1 - (x_D)_i]}$$

and similarly for component j:

$$(x_B)_j = \frac{(x_D)_j}{(x_D)_j + M[1 - (x_D)_j]}$$

Or still other arrangements or rearrangements may be formed.

Mole Fraction Summations

Note that the mole fraction summations are automatically satisfied. That is, since

$$(x_D)_i + (x_D)_j = 1$$

then

$$\left[K_i(x_F)_i - (x_F)_i \frac{L}{V} \right] + \left[K_j(x_F)_j - (x_F)_j \frac{L}{V} \right] = 1 - \frac{L}{V}$$

where

$$K_i(x_F)_i + K_j(x_F)_j = 1$$

This is but the bubble-point type calculation on the feedstream, which establishes V, since both K_i and K_j are functions of V. This is consistent with the assumptions that $\overline{V} = V$ and the feed is perfectly mixed with stream L_{m+1} and of the same composition.

Likewise, since

$$(x_B)_i + (x_B)_j = 1$$

then

$$\left[(x_F)_i - (x_F)_i \frac{\overline{V}K_i}{\overline{L}} \right] + \left[(x_F)_j - (x_F)_j \frac{\overline{V}K_j}{\overline{L}} \right] = 1 - \frac{\overline{V}}{\overline{L}}$$

where again

$$K_i(x_F)_i + K_j(x_F)_j = 1$$

The same consistency is verified.

Overall Material Balances

The overall material balances also are automatically satisfied. As before,

$$F = D + B$$

$$F(x_F)_i = D(x_D)_i + B(x_B)_i$$

$$F(x_F)_j = D(x_D)_j + B(x_B)_j$$

so that

$$\frac{B}{D} = \frac{(x_D)_i - (x_F)_i}{(x_F)_i - (x_B)_i} = \frac{(x_D)_j - (x_F)_j}{(x_F)_j - (x_B)_j}$$

Substituting,

$$\frac{(x_F)_i - \dfrac{K_i (x_F)_i}{1 - L/V}(1 - L/VK_i)}{\dfrac{(x_F)_i}{1 - \overline{V}/\overline{L}}(1 - \overline{V}K_i/\overline{L}) - (x_F)_i} = \frac{(x_F)_j - \dfrac{K_j (x_F)_j}{1 - L/V}(1 - L/VK_j)}{\dfrac{(x_F)_j}{1 - \overline{V}/\overline{L}}(1 - \overline{V}K_j/\overline{L}) - (x_F)_j}$$

Simplifying,

$$\frac{(1 - L/V) - K_i(1 - L/VK_i)}{(1 - \overline{V}K_i/\overline{L}) - (1 - \overline{V}/\overline{L})} = \frac{(1 - L/V) - K_j(1 - L/VK_j)}{(1 - \overline{V}K_j/\overline{L}) - (1 - \overline{V}/\overline{L})}$$

or

$$\frac{(1 - K_i)}{\dfrac{\overline{V}}{\overline{L}}(1 - K_i)} = \frac{(1 - K_j)}{\dfrac{\overline{V}}{\overline{L}}(1 - K_j)}$$

$$1 \equiv 1$$

Thus the overall material balances are satisfied.

EXAMPLE 4.1

The same arbitrary membrane characteristics and operating conditions are used as in the case of Example 3.1.

It was determined in the bubble-point type calculation for Example 3.1, where $V/F = 0$, that

$$V'' = 12.9317$$

$$K_i = 1.133536 \qquad K_j = 0.910977$$

$$1/K_i = 0.882195 \qquad 1/K_j = 1.097723$$

This determination is, first, for the purposes of assigning K-values for estimating the degree of separation. Other V/F ratios could have been used as well via the calculations of Example 3.1, but the bubble-point type calculation is the simplest. Moreover, the values of K do not change appreciably with the V/F ratio nor does the value of V''. It may be further added that, in the notation used here, when $X = 1$,

$$V'' = \overline{V}'' \qquad \overline{L} = L + F$$

That is to say, the feedstream mixture is assumed injected into the reject side of the membrane cell at the feed location. This is the simplest embodiment, since V'' does not change between the rectifying and stripping sections. It could as well be injected into the permeate side, however, or prorated between the reject and permeate sides.

The preceding is a parametric proration or partitioning of the feedstream as used, for instance, in the McCabe-Thiele method of distillation calculations or the Ponchon-Savarit method utilizing the H-x diagram, where the feedstream can be regarded as a saturated liquid, a saturated vapor, or a combination of the two, as previously noted.[1,2,4]

The use of a y-x curve in the McCabe-Thiele method for binary distillation calculations brings up the matter of a flash-vaporization representation, in case the feedstream mixture is at saturation. An inspection of the y-x curve relative to a given feed composition shows that the equilibrium mixture varies along the curve over a range from the bubble point (where the liquid phase composition x is equal to that of the feed mixture x_F) to the dew point (where the vapor composition y is equal to that of the feed mixture x_F). Between the two is the region of flash vaporization, where the equilibrium compositions (y, x) respectively of phases V and L must satisfy the flash material balance relation $F = V + L$, where

$$\frac{L}{V} = \frac{y - x_F}{x_F - x}$$

That is, dropping down vertically at constant x from a selected intermediate position on the K line, an arbitrary value for y can be selected. These values of x and y can then be used to determine the corresponding L/V ratio for the flash. The determination is not trial and error as presented. However, if the ratio L/V is preselected, then the determination becomes trial and error. In principle, there is the likelihood that the ratio L/V corresponds to the ratio $X/(1 - X)$ as previously derived, and any point on this latter line can be used to determine the flash compositions. Furthermore, these positions can be assigned as the compositions at the feed stage.

As previously mentioned, a problem to be avoided in all cases is that a zone of infinite plates can be encountered on both sides of the feed location if the intersection of the operating lines touches the K line. This defines a condition of minimum reflux or recycle.

Or we can speak of merely a feedstream, period. Here, we simply denote the feedstream as either injected into the reject side of the cell (as if a liquid), the permeate side (as if a vapor), or prorated or partitioned into both. Normally, assuming the feed mixture is injected into the reject side (as if a saturated liquid, in distillation) suffices. Accordingly, all stream sizes are calculated with respect to the permeate flux V''. This, in turn, is related to the actual stream flow rates (and membrane area). This information is utilized in the hand-calculated example, as follows.

Calculation for $n = 5$ and $m = 5$
The feed composition is

$$(x_F)_i = 0.4 \quad \text{and} \quad (x_F)_j = 0.6$$

Assuming $L/V = 0.4$ for the first trial,

$$A_i = (0.4)(0.882195) = 0.352878$$
$$A_j = (0.6)(1.097723) = 0.439089$$

and it follows that

$$1 - (A_i)^6 = 0.9980692 \qquad 1 - (A_j)^6 = 0.9928333$$
$$[1 - (A_i)^5](0.4) = \underline{0.3978113} \quad [1 - (A_j)^5](0.4) = \underline{0.3934714}$$

Difference $\qquad\qquad\qquad\qquad 0.6002579 \qquad\qquad\qquad\qquad 0.5993619$

Accordingly,

$$(x_D)_i = K_i (x_F)_i \frac{1 - A_i}{(\text{Difference})_i} = (1.133536)(0.4)\frac{1 - 0.352878}{0.6002579} = 0.488814$$

$$(x_D)_j = K_i (x_F)_j \frac{1 - A_j}{(\text{Difference})_j} = (0.910977)(0.6)\frac{1 - 0.439089}{0.5993619} = 0.511521$$

For the sum on the right,

$$\sum = 1.000335$$

which is sufficiently close for the purposes here. Note that, for a given value of n, this sum increases as L/V is increases.

For the stripping section, also assuming that here $\overline{V/L} = 0.4$ for the first trial,

$$S_i = 0.4(1.133536) = 0.453414$$

$$S_j = 0.4(0.910977) = 0.364391$$

$1 - (S_i)^6 = 0.991311$	$1 - (S_j)^6 = 0.997659$
$[1 - (S_i)^5](0.4) = \underline{0.392335}$	$[1 - (S_j)^5](0.4) = \underline{0.397430}$

Difference $\qquad\qquad\qquad 0.598976 \qquad\qquad\qquad\qquad\qquad 0.600229$

Accordingly,

$$(x_B)_i = (x_F)_i \frac{1 - S_i}{(\text{Difference})_i} = 0.4 \frac{1 - 0.453414}{0.598976} = 0.365014$$

$$(x_B)_j = (x_F)_j \frac{1 - S_j}{(\text{Difference})_j} = 0.6 \frac{1 - 0.364391}{0.600229} = 0.635367$$

For the sum on the right,

$$\sum = 1.00381$$

This is regarded as the solution. Note also that, for a given value of m, this sum increases as $\overline{V/L}$ increases.

It follows that

$$L'' = (L/V)V'' = (0.4)(12.9312) = 5.1725$$

$$D'' = V'' - L'' = 12.9312 - 5.1725 = 7.7587$$

$$\overline{L}'' = [1/(\overline{V/L})]\overline{V}'' = [1/0.4](12.9312) = 32.3280$$

$$B'' = \overline{L}'' - \overline{V}'' = 32.3280 - 12.9312 = 19.3968$$

In turn,

$$F'' = \overline{L}'' - L'' = 32.3280 - 5.17225 = 27.1555$$

$$= D'' + B'' = 7.7587 + 19.3968 = 27.1555$$

For the overall material balance for component i,

27.1555(0.4) vs. 7.7587(0.488814) + 19.3968(0.365013)

11.8622 vs. 3.79256 + 7.0801

10.8622 vs. 10.8726

For component j,

27.1555(0.6) vs. 7.7587(0.511520) + 19.3968(0.635367)

16.2933 vs. 3.9687 + 12.2241

16.2933 vs. 16.2928

These results are considered sufficiently close, in that the component material balances are largely met. There is the indication that the component material balances are automatically satisfied, as previously pointed out.

It may also be noted that $\overline{V}/V = 12.9312/12.9312 = 1$ and $\overline{L}/L = 32.3280/5.1725 = 6.2500$. Note furthermore that

$$V/F = 12.9312/27.1555 = 0.4762$$
$$\overline{V}/F = 12.9312/27.1555 = 0.4762$$

This ratio can be utilized to prorate the permeate flux to the feed rate and, in turn, to membrane cell area per mole of feedstream, as shown in Example 3.1.

The foregoing results apply to $n = 5$ and $m = 5$. For different values of n and m, different results expectedly are attained. Predictably, as n and m increase, the sharpness of separation is enhanced. It may also be commented that the results are relatively insensitive to variations in L/V and $\overline{V}/L$. Generally speaking, however, an increase in each increases the degree of separation.

The corresponding spreadsheet calculations are furnished in Appendix 4, including the determination of membrane cell area, and these may be applied to other circumstances and specifications. More rigorously speaking, there is a relationship between the rectifying section and stripping sections, which, given L/D or L/V and x_D, implies a relationship between $\overline{V}/B$ or $\overline{L}/V$ (or $\overline{V}/L$) and x_B, as previously derived for $X = 1$. This, however, introduces an extra element of trial and error, a refinement not always warranted for estimation purposes.

Simplified Calculations

For sufficiently large values of n and m and sufficiently small values of A and S, the approximation for the rectifying section yields

$$(x_D)_i = \frac{K_i - \dfrac{L}{V}}{1 - \dfrac{L}{V}} = \frac{1.133536 - \dfrac{L}{V}}{1 - \dfrac{L}{V}}(0.4)$$

$$(x_D)_j = \frac{K_j - \dfrac{L}{V}}{1 - \dfrac{L}{V}} = \frac{1.133536 - \dfrac{L}{V}}{1 - \dfrac{L}{V}}(0.4)$$

A comparison is as follows:

L/V	$(x_D)_i$	$(x_D)_j$	Σ
0.4	0.489024	0.510977	1.00000
0.6	0.533536	0.466455	1.00000
0.8	0.667072	0.332931	1.00000
0.85	0.756096	0.243908	1.00000

As L/V increases, the sharpness of separation increases. Note that for $L/V = 0.4$, the results are similar to the case for $n = 5$.

For the stripping section,

$$(x_B)_i = \frac{1 - \dfrac{\overline{V}}{\overline{L}}K_i}{1 - \dfrac{\overline{V}}{\overline{L}}}(x_F)_i = \frac{1 - \dfrac{\overline{V}}{\overline{L}}(1.133536)}{1 - \dfrac{\overline{V}}{\overline{L}}}(0.4)$$

$$(x_B)_j = \frac{1 - \dfrac{\overline{V}}{\overline{L}}K_j}{1 - \dfrac{\overline{V}}{\overline{L}}}(x_F)_j = \frac{1 - \dfrac{\overline{V}}{\overline{L}}(0.910977)}{1 - \dfrac{\overline{V}}{\overline{L}}}(0.6)$$

A comparison is as follows:

$\overline{V}/\overline{L}$	$(x_B)_i$	$(x_B)_j$	Σ
0.4	0.364390	0.635609	1.0000
0.5	0.346586	0.653414	1.0000
0.6	0.319878	0.680121	1.0000
0.7	0.275366	0.724632	1.0000
0.8	0.186342	0.813655	1.0000
0.85	0.097318	0.902678	1.0000

As $\overline{V}/\overline{L}$ increases, the sharpness of separation increases. Note that, for $\overline{V}/\overline{L} = 0.4$, the results are similar to those for $m = 5$. In each case, the component material balances are satisfied, as demonstrated.

4.12   CONCLUSIONS

Flash vaporization and multistage distillation calculation methods have been shown to be adaptable to membrane separations, which indicates the degree of separation that can be achieved. Furthermore, the use of multistage or cascade operations enhances the sharpness of separation and can be used to separate components with low relative selectivity.

REFERENCES

1. Hoffman, E. J. *Azeotropic and Extractive Distillation.* New York: Wiley-Interscience, 1964; Huntington, NY: Krieger, 1977.
2. Brown, G. G., A. S. Foust, D. L. Katz, R. Schneidewind, R. R. White, W. P. Wood, G. M. Brown, L. E. Brownell, J. J. Martin, G. B. Williams, J. T. Banchero, and J. L. York. *Unit Operations.* New York: Wiley, 1950.
3. Lee, S. Y., and B. S. Minhas. "Effect of Gas Composition and Pressure on Permeation through Cellulose Acetate Membranes." In *New Membrane Materials and Processes for Separation,* ed. Kamalesh K. Sirkar and Douglas R. Lloyd. AIChE Symposium Series, vol. 84, no. 261, 1987 AIChE Summer National Meeting, Minneapolis. New York: American Institute of Chemical Engineers, 1988.
4. Badger, W. L., and W. L. McCabe. *Elements of Chemical Engineering.* New York: McGraw-Hill, 1936.
5. Katz, D. L., D. Cornell, R. Kobayashi, F. H. Poettmann, J. A. Vary, J. R. Elenbaas, and C. F. Weinaug. *Handbook of Natural Gas Engineering.* New York: McGraw-Hill, 1959.
6. Hoffman, E. J. *Phase and Flow Behavior in Petroleum Production.* Laramie, WY: Energon, 1981.
7. Souders, Mott, Jr., and George Granger Brown. "IV. Fundamental Design of Absorbing and Stripping Columns for Complex Vapors." *Industrial Engineering Chemistry* 24 (1932), pp. 519–522.

5

Differential Permeation with Point Permeate Withdrawal

In differential permeation, the point compositions of the phases are considered to vary linearly with position along the surface(s) of the membrane. A steady state is assumed, so that the compositions are independent of time.

One phase is called the permeate V, the other phase the reject L. The designators also refer to the (molar) flow rate of each of the phases, which varies with position. The reject is the continuation of the feed-stream. Its inlet rate and composition are those of the feed, its final or exit rate and composition depend on the permeation effected.

If the permeate is withdrawn at each point along the membrane surface, then the situation corresponds somewhat to that of bulk differential vaporization or condensation.[1,2] However, in the case at hand, the reject is considered to be a flow system of changing composition rather than a bulk phase of uniform composition. Moreover, the change in composition of the reject phase is to be a function of linear position only and not of time.

The permeate so withdrawn may subsequently be accumulated, but this is independent of the permeation process per se. Furthermore, the accumulation may be withdrawn as if concurrent to the flow of the reject phase or as if countercurrent to the reject phase. The total accumulation and its composition are the same in each situation; it is "after the fact."

5.1   DIFFERENTIAL PERMEATION

For membrane systems in concurrent or countercurrent flow of the reject and permeate, the concept of differential permeation applies. Here we consider the flow of the feed and reject stream as diagrammed in Figure 5.1. The end of the membrane cell where the feed stream is introduced

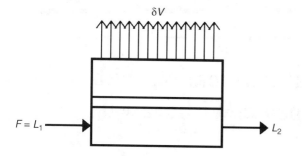

Figure 5.1
Differential
permeation with point
permeate withdrawal.

is designated 1, the other end is designated 2. Stream L_1 corresponds to the feed F, as previously used. Stream L is the reject at any point, and stream L_2 is the final reject.

The accumulated permeate stream V may move concurrently with or countercurrently to stream L. In the configuration for perfect mixing shown in Figure 3.1, for comparison, none of the permeate is withdrawn intermediately, say, between 1 and 2 or between 2 and 1, as is indicated in Figure 5.1.

At each point along the membrane, a quantity δV of permeate of composition y_i is passed through the membrane. In this respect, it is similar to differential vaporization or condensation.

Furthermore, in concurrent flow, V_1 equals zero as a limiting condition, even though it has initial composition values, to be determined. In countercurrent flow, V_2 may equal zero, even though it has composition values, to be determined.

Bubble-Point Type Curve

In all cases, stream L follows the bubble-point equivalent curve, here embodied as the K line, as in differential vaporization.[2] The stream V represents the accumulation of permeate and may include any stream injected as V_1 in concurrent flow or V_2 in countercurrent flow.

5.2 OVERALL MATERIAL BALANCES

The overall material balances are as follows for the different juxtapositions.

Concurrent Flow

Here,

$$V_1 + L_1 = V + L = V_2 + L_2$$

and for any component i, dropping the component subscript,

$$V_1 y_1 + L_1 x_1 = Vy + Lx = V_2 y_2 + L_2 x_2$$

where, as a special case, $V_1 = 0$. The separation and recovery, in turn, follow. Given L_1 and V_1 and assuming L_2, say, will affix V_2.

Countercurrent Flow

In this case,

$$V_1 - L_1 = V - L = V_2 - L_2$$

and for any component i,

$$L_1 x_1 - V_1 y_1 = Lx - Vy = L_2 y_2 - V_2 y_2$$

As a special case, $V_2 = 0$.

It may be observed that, given L_1, affixing V_2 provides the limit V_1 in terms of the limit L_2 or vice versa.

Semi-Continuous or Semi-Batch Flow

In this embodiment, it is assumed that the reject stream L may flow in either direction and be depleted and that the permeate phase V accumulates "in bulk." In other words, perfect mixing of the permeate occurs. For the limited purposes of this chapter, this is the configuration studied.

5.3   DIFFERENTIAL MATERIAL BALANCES

Differential material balances can be written based on a "drop" of permeate passing through the membrane and thus depleting the reject phase L. Therefore, for semi-continuous flow,

$$-dL = dV$$
$$-d(Lx_i) = y_i dV$$

where $dV = \delta V$ represents the "drop" of permeate that passes through the membrane.

It follows that

$$-L dx_i = (y_i - x_i) dV$$

or, on reintroducing component subscripts,

$$-\frac{dL}{L} = -d \ln L = \frac{dx_i}{x_i - y_i} = \frac{dx_i}{x_i - K_i x_i} = \frac{1}{1 - K_i} \frac{dx_i}{x_i}$$

where $\Sigma x_i = 1$ or $\Sigma dx_i = 0$. Note that the K_i are functions of the flux L'' (or V''), which introduces the membrane area. In principle, this relates V to x_i (or y_i). Numerical integration procedures are required, however, which can prove exceedingly complex for a multicomponent system due to the summation constraints for the mole fractions.

For a two-component system i and j, however,

$$\frac{dx_i}{dx_j} = -1 = \frac{y_i - x_i}{y_j - x_j} = \frac{x_i(K_i - 1)}{x_j(K_j - 1)}$$

or

$$-K_j x_j + x_j = K_i x_i - x_i$$

or

$$x_i + x_j = 1 = K_i x_i + K_j x_j$$

which is but the bubble-point type calculation for a two-component or binary system.

It may be alternately derived that, for the two components i and j,

$$(-d \ln L)(x_i - K_i x_i) = dx_i$$
$$(-d \ln L)(x_j - K_j x_j) = dx_j$$

Taking the sum, since $dx_i + dx_j = 0$, and rearranging,

$$x_i + x_j = 1 = K_i x_i + K_j x_j$$

This is again but the bubble-point type of calculation, so the relationships are consistent.

Numerical Integration

The previously derived expression

$$-d \ln L = \frac{1}{x_i - y_i} dx_i$$

can be integrated to yield the form

$$L_2/L_1 = \exp\left(-\int_1^2 \frac{1}{x_i - y_i} dx_i\right)$$

or

$$L_2/L_1 = 1 \bigg/ \exp\left(\int_1^2 \frac{1}{x_i - y_i} dx_i\right)$$

where integration is between any two arbitrary points 1 and 2.

Expressed in the terms of numerical integration,

$$L_2/L_1 = 1 \bigg/ \exp\left(\sum_1^2 \frac{1}{x_i - y_i} \Delta x_i\right)$$

where the summations denote a succession of partial sums as point 2 takes on a succession of values relative to point 1. The exercise is performed in Table 5.2 of Example 5.1 and is subsequently shown in Appendix 5 using spreadsheet calculations.

5.4 BUBBLE-POINT TYPE CALCULATION

The bubble-point type determination at each point is, as noted,

$$K_i x_i + K_j x_j = 1$$

where $x_i + x_j = 1$ or $x_i = 1 - x_j$.

As previously determined, K_i is a function of the permeation characteristics; that is, for two components i and j, in the notation for permeation,

$$K_i = \frac{b}{V'' + a}$$

$$K_j = \frac{d}{V'' + c}$$

where V'' is the point permeation flux, that is, is the permeation rate per unit area at a point. The units are to be consistent; in other words, consistent with a, b, c, and d or vice versa. And as derived in Chapter 3,

$$a = P_i P_V$$
$$b = P_i P_L$$
$$c = P_j P_V$$
$$d = P_j P_L$$

Permeability therefore pertains to a unit area, or possibly a unit length, but would still be on an areal basis. In the notation conventions used for heat and fluid flow, it could be stated that $V'' = \delta V/\delta A$, where A is area. Alternately, it could be written that $V'' = \delta V/(A/s)\delta s$, where (A/s) stands for area per unit length. (Note that the superscript is double prime.) Furthermore, V'' is a function of composition, where

$$V'' = \frac{-B \pm \sqrt{B^2 - 4AC}}{2A}$$

and here, in the appropriate notation as used in Section 3.4,

$$A = 1$$

$$B = (\alpha + \beta) - [\bar{x}_i + \bar{x}_j]$$

$$C = (-\beta\bar{x}_i + \alpha\bar{x}_j) + \alpha\beta$$

whereby, for the bubble-point point type calculation on stream L—that is, at $V/F \rightarrow 0$—the quantities used above reduce to

$$\alpha = a$$

$$\beta = c$$

$$\bar{x}_i = bx_i$$

$$\bar{x}_j = dx_j$$

where $x_i + x_j = 1$ and where here d denotes the defined quantity and d the differential operator. The mole fractions x_i and x_j vary between the limits assigned for the flow of the reject phase L.

5.5 ACCUMULATION

The total accumulation n_T of the permeate phase V is given by

$$n_T = \int_1^2 dV = -\int_1^2 dL$$

The accumulation of component i in the permeate is

$$(n_T)_i = \int_1^2 y_i \, dV = -\int_1^2 y_i \, dL$$

where y_i is the composition of a drop of permeate δV. A similar equation can be written for component j.

5.6 DIFFERENTIAL RATE BALANCES

The rate balance for transferring the totality of both components of a binary mixture simultaneously may be written as

$$(y_i + y_j)dV = \{P_i(P_L x_i - P_V y_i) + P_j(P_L x_j - P_V y_j)\}dA$$

or, since the mole fractions sum to unity and $dL = -dV$, also dividing by L_1 as a convenience,

$$d(A/L_1) = \{P_i(P_L x_i - P_V y_i) + P_j(P_L x_j - P_V y_j)\}^{-1}(-)d(L/L_1)$$

This expression gives the membrane interfacial area as a function of L (or as a function of y_i or x_i, since all are related by the material balances).

5.7 EQUILIBRIUM

If the total rate of mass transfer should become zero, then it is required that

$$P_i(P_L x_i - P_V y_i) + P_j(P_L x_j - P_V y_j) = 0$$

Since the mole fractions in each phase sum to unity,

$$P_i P_L x_i - P_i P_V y_i = -P_j P_L + P_j P_L x_i + P_j P_V - P_j P_V y_i$$

or

$$x_i[P_i P_L - P_j P_L] = y_i[P_i P_V - P_j P_V] - P_j(P_L - P_V)$$

or

$$x_i P_L(P_i - P_j) = y_i P_V(P_i - P_j) - P_j(P_L - P_V)$$

or, on solving for y_i,

$$y_i = x_i \frac{P_L}{P_V} + \frac{P_j}{P_i - P_j} \frac{P_V - P_L}{P_V}$$

Alternately, it can be shown that

$$y_j = x_j \frac{P_L}{P_V} + \frac{P_i}{P_i - P_j} \frac{P_V - P_L}{P_V}$$

These relationships, however, require that the components i and j have a net transfer in opposite directions. It can be considered as sort of a quasi-equilibrium or dynamic equilibrium.

If there is no net transfer of either component, then

$$P_L x_i - P_V y_i = 0$$

$$P_L x_j - P_V y_j = 0$$

These may be regarded as the conditions for a "true" equilibrium, albeit the concept of phase equilibrium, say, requires that the pressure be uniform throughout the system.

If the former condition holds, then

$$\frac{y_i}{x_i} = \frac{P_L}{P_V} = \frac{1 - y_j}{1 - x_j}$$

where

$$\frac{P_L}{P_V} - x_j \frac{P_L}{P_V} = 1 - y_j$$

or

$$y_j = (x_j - 1)\frac{P_L}{P_V} + 1 = -x_i \frac{P_L}{P_V} + 1$$

Therefore,

$$P_L x_j - P_V y_j = P_L (1 - x_i) - P_V \left(-x_i \frac{P_L}{P_V} \right)$$

$$= P_L - P_V$$

Whereas the second condition for "true" equilibrium requires that

$$P_L x_j - P_V y_j = 0$$

and a contradiction ensues. It follows, therefore, that true equilibrium can exist only if $P_V = P_L$ and $y_j = x_j$ (and $y_i = x_i$). These are the normal expectations for a thermodynamic equilibrium to occur.

EXAMPLE 5.1

A membrane separation is to be conducted with a continuous withdrawal of the permeate. At each linear point along the membrane during the separation, the reject phase composition is governed by a bubble-point type determination, starting with the initial feed composition.

The data from Example 3.1 (which is based on Examples 2.1 and 2.2) is utilized and adapted as follows:

$$P_i = 20 \text{ g-moles of } i \text{ per cm}^2\text{-sec-atm}$$
$$P_j = 10 \text{ g-moles of } i \text{ per cm}^2\text{-sec-atm}$$
$$P_L = 3(10^1) \text{ atm}$$
$$P_V = 2(10^1) \text{ atm}$$
$$K_i = 60/(V'' + 40)$$
$$K_j = 30/(V'' + 20)$$
$$(x_1)_i = 0.4$$
$$(x_1)_j = 0.6$$

Furthermore,

$$b = 60 \qquad a = 40$$
$$d = 30 \qquad c = 20$$

and, for $V/F = 0$,

$$\alpha = 40$$
$$\beta = 20$$
$$(\bar{x}_1)_i = 60x_i$$
$$(\bar{x}_1)_j = 30x_j$$

In turn,

$$V'' = \frac{-B \pm \sqrt{B^2 - 4AC}}{2A}$$

where

$$A = 1$$
$$B = (40 + 20) - [60x_i + 30x_j]$$
$$= 60 - [60x_i + 30(1 - x_i)]$$
$$= 30 - 30x_i = 30(1 - x_i) = 30x_j$$
$$C = -[20x_i + 40x_j] + 800$$
$$= -[20(60)x_i + 40(30)x_j] + 800$$
$$= -1200(x_i + x_j) + 800 = -400$$

Table 5.1 Determination of Constants

x_i	x_j	$\bar{x}_i$	$\bar{x}_j$	B	$\beta\bar{x}_i$	$\alpha\bar{x}_j$	C	V''
0.408	0.60	24	18	18	480	720	−400	12.9317
0.39	0.61	23.4	18.3	18.3	468	732	−400	12.8435
0.35	0.65	21	19.5	19.5	420	780	−400	12.5000
0.30	0.70	18	21	21	360	840	−400	12.0887
0.20	0.80	12	24	24	240	960	−400	11.3238
0.10	0.90	6	27	27	120	1080	−400	10.6299
0.05	0.95	3	28.5	28.5	60	1140	−400	10.3073
0.01	0.99	0.6	29.7	29.7	12	1188	−400	10.0603
0.00	1.00	0	30	30	0	1200	−400	10.0000

The calculations are tabulated in Tables 5.1 through 5.3. Note that, from the results of Table 5.3, the interfacial area requirement for the membrane varies almost uniformly with the change in the amount of the reject phase (or change in the amount of the permeate phase).

The total interfacial area requirement may be estimated from the summation presented in Table 5.3, as follows:

$$0.07760(0.01) = 0.000776$$
$$0.07893(0.04) = 0.0031572$$
$$0.08137(0.05) = 0.0040685$$
$$0.08552(0.1) = 0.008552$$
$$0.09119(0.1) = 0.009119$$
$$0.09554(0.05) = 0.004777$$
$$0.09821(0.04) = 0.0039284$$
$$0.0997(0.01) = \underline{0.000997}$$
$$\text{Total} \quad 0.0353751$$

The answer is as yet dimensionless.

Determination of Membrane Area

A rough determination for the required membrane area in dimensionless units may be obtained by averaging the values of dA/dV in the last column of Table 5.3 and multiplying by unity; that is, all the feed is assumed converted to permeate. The averaged value is 0.08851. (It may be observed that the derivative dA/dV remains relatively constant.)

Table 5.2 Determination of the Incremental Relative Reject Rate

x_i	K_i	y_i	y_i	$(x_i - y_i)^{-1}$	$\dfrac{\Delta x_i}{(x_i - y_i)_{av}}$	$\dfrac{\Delta x_i}{(x_i - y_i)_{av}}$	L/L_1
0.40	1.133536	0.4534	0.5466	−18.73	0.1884	0.0000	1.00000
0.39	1.135428	0.4428	0.5572	−18.94	0.7788	0.1884	0.82828
0.35	1.142857	0.4000	0.6000	−20.00	1.0483	0.9672	0.38015
0.30	1.151881	0.3456	0.6544	−21.93	2.5760	2.0155	0.13325
0.20	1.169048	0.2338	0.7662	−29.59	4.182	4.5915	0.01014
0.10	1.185070	0.1185	0.8815	−54.05	3.956	8.774	0.00015
0.05	1.192670	0.0596	0.9404	−104.17	12.083	12.730	0.000003
0.01	1.198555	0.0120	0.9880	−500.00	—	24.813	1.7×10^{-11}
0.00	1.200000	0.0	1.0000	—	—	—	0.0

Table 5.3 Determination of $dA/dV = -dA/dL$

x_i	$P_L x_i$	$P_V y_i$	P_i (difference)	$P_L x_i$	$P_V y_i$	P_i (difference)	Sum^{-1}	Average
0.4	1.2	0.9068	5.8640	1.8	1.0932	7.0680	0.07733	0.07760
0.39	1.17	0.08856	5.6880	1.83	1.1144	7.1560	0.07786	0.07893
0.35	1.05	0.80000	5.0000	1.95	1.2000	7.5000	0.08000	0.08137
0.3	0.9	0.6912	4.1760	2.1	1.3088	7.912	0.08273	0.08552
0.2	0.6	0.4676	2.6480	2.4	1.5324	8.676	0.08831	0.09119
0.1	0.3	0.2370	1.260	2.7	1.763	9.370	0.09407	0.09554
0.05	0.15	0.1192	0.616	2.85	1.8808	9.692	0.09701	0.09821
0.01	0.03	0.0240	0.120	2.97	1.976	9.940	0.09940	0.0997
0.00	0.00	0.0000	0.000	3.00	2.000	10.000	0.10000	

In turn, if the values for the permeability are in the units of

$$10^{-9} \text{ cm}^3/\text{cm}^2\text{-sec-cm Hg/cm}$$

where the membrane thickness is to be 10 microns or $10(10^{-4})$ cm and the pressures are in (10^1) atm, then the conversion factor on a molar basis is, as determined elsewhere,

$$\frac{10^{-9}(76/22,414)(10)}{10(10^{-4})} = 0.0339(10^{-6})$$

Accordingly, the actual areal requirement on this basis is

$$A = \frac{0.08851}{0.0339(10^{-6})} = 7.26(10^6) \text{ cm}^2$$

for a feed rate of 1 gram-mole per sec and assuming all the feed is transferred to permeate. (It may be added that 929 cm^2 = 1 ft^2.) A more rigorous spreadsheet determination is furnished in Appendix 5. The above is as compared to $A = 1.16(10^6)$ cm^2 in Examples 1.2 and 2.2.

REFERENCES

1. Hoffman, E. J. *Azeotropic and Extractive Distillation*. New York: Wiley-Interscience, 1964; Huntington, NY: Krieger, 1977.
2. Hoffman, E. J. *Heat Transfer Rate Analysis*, pp. 380ff. Tulsa, OK: PennWell, 1980.

6

Differential Permeation
with Permeate Flow

For this circumstance, flow is regarded as parallel to the inner and outer membrane surfaces, and the flow of each phase is regarded as concurrent with or countercurrent to the other phase. Furthermore, we are speaking of what is regarded as "plug flow"; that is, no forward or backward mixing. The juxtaposition is diagrammed in Figure 6.1. One phase, L, is the reject; the other phase, V, is the permeate. The reject is the continuation of the feedstream.

Perhaps the simplest embodiment of a membrane cell is tubular, with the tubeside flow concurrent or countercurrent to the flow outside the tube, as described in Chapter 1. The tube may be positioned inside another tube, called a *tube in a tube*, so that this latter flow is within the annulus. Such an arrangement is diagrammed schematically in Figure 6.2.

Alternately, the cell may be designed similarly to a shell-and-tube heat exchanger, with flow inside the tubes and on the outside or shell side. The shell-side flow may be strictly parallel to the tubes or also across the tubes, or tube bundle, and directed by the use of baffles and baffle cuts. Such a layout is illustrated in Figure 6.3, with more information about the intricacies provided by Hoffman.[1] There is an analogy with the treatment of absorbers, strippers, and distillation columns as a continuum, described in terms of the rate of mass transfer.[2]

Perhaps the simplest embodiment, at least for concurrent flow, is to regard the flow system as a case of perfect mixing at each point of a continuum, but where the composition varies from point to point of the continuum, as well as the stream rates, from one end to the other. As an example, at point 1, the composition of the feed or initial reject stream $(F = L_1)$ could be said to be at its "bubble point." Furthermore, the bubble-point composition as calculated for V is the composition for the first drop of V produced; that is to say, in this particular embodiment, $V_1 = 0$.

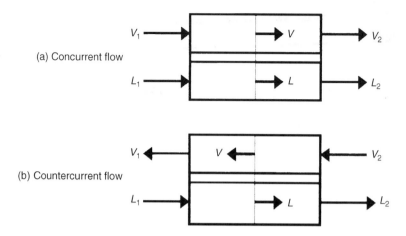

Figure 6.1 Differential permeation as a continuum.

Figure 6.2 Single-pass tube-in-tube heat exchanger.

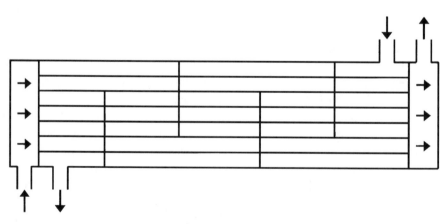

Figure 6.3 Single-pass shell-and-tube heat exchanger.

In turn, as the ratio V/F (or L/F) varies as shown in Example 3.1, the value of V so calculated can more appropriately described as a flux at each particular point and can be based on some arbitrary value of F, where the molar flux designator V''' is more appropriate. That is, for the purposes here, $V \equiv V'''$, with the dimensions of overall permeability times pressure, say, in moles per unit area per unit time (or whatever is to be chosen, as per Section 2.1 and the examples of Chapter 2).

For a complete cocurrent transfer of material, the determination at point 2 represents the "dew-point" at $L_2 = 0$ of $V_2 = F = L_1$. Alternately, it can be perceived as the "bubble point" of L_2; that is, where $V_2 = F = L_1$ and $L_2 \to 0$. Note, furthermore, that for this particular case (for total transfer), the composition of V_2 would become identical to that of $F = L_1$, as would the stream size. In other words, a complete transfer of feed material has occurred from stream L to stream V.

It may be observed that, at a point, $V'' = dV/dA$ or $dA = dV/V''$, where V denotes the total stream flow rate, say, in moles per unit time. Integrating between points 1 and 2,

$$A = \int_1^2 \frac{1}{V''} dV \sim \sum_1^2 \frac{1}{V''} \Delta V$$

From Table 3.2 of Example 3.1, note that V'' (or V) varies only slightly between the values $V/F = 0$ and $V/F = 1$: The range is from 12.9317 to 12.500. Merely using an average value of 12.72 for simplicity, the integral or summation calculates readily to

$$A \sim (1/12.72)(1) = 0.0786$$

in the arbitrary units used for Example 3.1. This is close to the same result as would be obtained in Example 3.1 for perfect mixing, where $V/F \sim 0.467$.

In Example 3.1, the membrane thickness is not specified. Let this value be 10 microns or $10(10^{-4})$ cm. Using the same units for permeability (cc^3/cm^2-sec-cm Hg/cm) and pressure (10^1 atm) that were prescribed at the end of Example 3.1, the conversion factor for V'' (or V) is

$$[(10^{-9})(76/22,414)(10)]/[10(10^{-4})] = [76/22,414](10^{-5}) = 0.0339(10^{-6})$$

Accordingly, the total required membrane area would be

$$0.0786/[(76/22,414)(10^{-5})] = 2.318(10^6) \ cm^2$$

per g-mole of feed per sec, where $929 \ cm^2 = 1 \ ft^2$.

Interestingly, this figure is about twice the area calculated for Example 3.1 for $V/F = 0.5$, and in the corresponding spreadsheet calculations of Appendix 3. It is also similar to the area calculated in Chapter 5, as per the spreadsheet calculations of Appendix 5. The value is likewise similar to the area determined via the spreadsheet calculations as presented in Appendix 6. All this similarity can be viewed as substantiating the concept of single-stage perfect mixing as originally developed and utilized in Chapter 3.

6.1 MATERIAL AND RATE BALANCES

The material and rate balances are negative or positive, depending on whether flow is concurrent or countercurrent. As set forth in Figure 6.1, the convention used is that the direction of integration is from one end of the cell to the other for each stream; that is, the limits for V and L are at the same common ends of the cell, designated point 1 and point 2, albeit the flows are in opposite directions in the case of countercurrent flow.

Concurrent Flow

The differential material balances read as follows:

$$dV = -dL$$

$$d(Vy_i) = -d(Lx_i)$$

The differential rate balance for each component i is

$$d(Vy_i) = P_i(P_L x_i - P_V y_i)dA$$

where A is the membrane interfacial surface area and the units are to be consistent. The convention is that the permeate V increases as A increases (that is, dA is positive in the direction of integration). At the same time, the reject phase L decreases. The amount of component i in the permeate (Vy_i) also increases in the direction of integration, as does the amount of component i in the reject (Lx_i). The interfacial area of the membrane material per se may be based on the inner or outer surface, as specified in the permeability determination.

Countercurrent Flow

Here, the positive sign is introduced such that

$$dV = dL$$

$$d(Vy_i) = d(Lx_i)$$

The differential rate balance for each component i involves a negative sign, however, and is, in this case,

$$-\mathrm{d}(Vy_i) = P_i(P_L x_i - P_V y_i)\mathrm{d}A$$

The convention is that, when the term on the right has a positive value and $\mathrm{d}A$ is positive in the direction of integration, then the permeate phase V decreases, as does the reject phase L, and similarly for the amount of component i, designated (Vy_i) for the permeate and (Lx_i) for the reject.

Solution

A solution requires establishing the necessary and sufficient relationships between y_i, x_i, and V or L, where the rate equation can be numerically integrated, at the same time yielding the behavior of A. This involves assigning or establishing the boundary conditions. Additional considerations are as follows. The nature of the set of equations is such that trial-and-error numerical stepwise procedures are required. Furthermore, the complexity is such that even the solution for a two-component system is a formidable undertaking.

In comparison, it may be observed that the determination of interfacial membrane area is not a consideration in determining the degree of separation for perfect mixing. The interfacial area is, in effect, built into the permeability coefficient for both situations, for single-stage and multistage separations. Alternately stated, the degree of separation for single-stage and multistage separations can be construed as based on a unit area of the membrane surface(s).

For the point withdrawal of the permeate as a bulk phase, the membrane or interfacial area can be determined after the fact; that is, it is not necessary to include A as a variable in establishing the a priori relationships among y_i, x_i, and L or V.

Boundary Conditions

In all cases, L_1 and its composition $(x_2)_i$ are specified, at point 1. Stream L_1 is identical to the feedstream F, as otherwise used. Furthermore, at point 1, the convention used is that A or $A_1 = 0$.

In concurrent flow, V_1 and its composition $(y_1)_i$ may also be specified. If $V_1 = 0$, its composition may be determined from a bubble-point type calculation on L_1; that is, the composition of V_1 is that of the first "drop" of permeate produced. Simultaneous numerical integrations can, in principle, be carried out to a specified value for any of the variables at point 2.

In countercurrent flow, V_2 and its composition $(y_2)_i$ may be specified. If $V_2 = 0$, then its composition necessarily is determined from a bubble-point type calculation on L_2. Unfortunately, neither L_2 nor its composition $(x_2)_i$ would yet be known. This introduces an extra or double trial-and-error element into the solution, even for a two-component system, where, say,

$$(x_2)_i = (x_2)_i - 1$$

and it is necessary to carry only one component through the calculations.

6.2 COMPONENT RELATIONSHIPS

The relationships between composition and flow rates can be set forth, based on the rate equations, for both concurrent and countercurrent flow, and based either on the permeate phase V or the reject phase L. The boundary conditions may make one preferable to the other.

Concurrent Flow

The rate equations for a two-component system are compared as follows, in consistent units:

$$\frac{d(Vy_i)}{P_i(P_L x_i - P_V y_i)} = \frac{d(Vy_j)}{P_j(P_L x_j - P_V y_j)} = dA$$

or

$$\frac{V dy_i + y_i dV}{P_i(P_L x_i - P_V y_i)} = \frac{-V dy_i + (1 - y_i)dV}{P_j[P_L(1 - x_i) - P_V(1 - y_i)]} = dA$$

or

$$\frac{V\dfrac{dy_i}{dV} + y_i}{\Phi_i} = \frac{-V\dfrac{dy_i}{dV} + (1 - y_i)}{\Phi_j} = \frac{dA}{dV}$$

where Φ_i and Φ_j are defined by the substitutions.
It may be observed also that

$$d(Vy_i) = d[V(1 - y_j)] = dV - d(Vy_j)$$

or

$$d(Vy_i) + d(Vy_j) = dV$$

Therefore, in general, $d(Vy_i) \neq -d(Vy_j)$, albeit $dy_i = -dy_j$.

On substituting for $d(Vy_j)$ in the rate equation, it follows that

$$\frac{d(Vy_i)}{\Phi_i} = \frac{-d(Vy_i) + dV}{\Phi_j}$$

On combining the fractions,

$$\frac{d(Vy_i)}{\Phi_i} + \frac{d(Vy_i)}{\Phi_j} = \frac{dV}{\Phi_j}$$

On clearing the fractions,

$$\frac{\Phi_j d(Vy_i) + \Phi_i d(Vy_i)}{\Phi_i \Phi_j} = \frac{dV}{\Phi_j}$$

whereby, on transposing i and j,

$$d(Vy_i) = \frac{\Phi_i}{\Phi_i + \Phi_j} dV$$

$$d(Vy_j) = \frac{\Phi_j}{\Phi_i + \Phi_j} dV$$

Furthermore, since

$$\frac{d(Vy_i)}{\Phi_i} = dA \qquad \text{and} \qquad \frac{d(Vy_j)}{\Phi_j} = dA$$

then

$$d(Vy_i) + d(Vy_j) = (\Phi_i + \Phi_j)dA$$

$$\frac{\Phi_i}{\Phi_i + \Phi_j} dV + \frac{\Phi_j}{\Phi_i + \Phi_j} dV = (\Phi_i + \Phi_j)dA$$

where

$$\frac{1}{\Phi_i + \Phi_j} dV = dA$$

Knowing the behavior of Φ_i and Φ_j, this is the most ready route for estimating incremental changes in A given the incremental change in V.

If $dV/dA = 0$, then $\Phi_i + \Phi_j = 0$. Furthermore, if $d(Vy_i)/dA = 0$, then $\Phi_i = 0$. Or, if for component j, on the other hand, $d(Vy_j)/dA = 0$, then $\Phi_j = 0$.

Determination of Composition Change in Terms of V

On cross-multiplying the rate balance between i and j,

$$\Phi_j \frac{V dy_i}{dV} + \Phi_j y_i = \Phi_i \left(-\frac{V dy_i}{dV} \right) + \Phi_i (1 - y_i)$$

or

$$V \frac{dy_i}{dV} (\Phi_i + \Phi_j) = -y_i (\Phi_i + \Phi_j) + \Phi_i$$

where, on introducing the feed rate F and using incremental notation,

$$\frac{\delta y_i}{\delta (V/F)} = \frac{F}{V} \left[-y_i + \frac{\Phi_i}{\Phi_i + \Phi_j} \right]$$

$$\frac{\delta y_i}{\delta (V/F)} = \frac{F}{V} \left[-y_i + \frac{\Phi_i}{\Phi_i + \Phi_j} \right]$$

or

$$\frac{\delta y_i}{\delta \ln (V/F)} = -y_i + \frac{\Phi_i}{\Phi_i + \Phi_j}$$

where, as defined and used before,

$$\Phi_i = P_i (P_L x_i - P_V y_i)$$
$$\Phi_j = P_j [P_L (1 - x_i) - P_V (1 - y_i)]$$

In this fashion, the incremental change in y_i can be determined in terms of an incremental change in V or V/F.

Note that, since $\delta y_i + \delta y_j = 0$, it follows that

$$0 = -y_i + \frac{\Phi_i}{\Phi_i + \Phi_j} - y_j + \frac{\Phi_j}{\Phi_i + \Phi_j}$$

Hence,

$$y_i + y_j = \frac{\Phi_i}{\Phi_i + \Phi_j} + \frac{\Phi_j}{\Phi_i + \Phi_j} = 1$$

and the previously derived expression for δy_i is consistent with the mole fraction balance. Furthermore, observe that $\delta y_i = -\delta y_j$, so that, as the mole fraction of say component i increases, that of component j decreases and vice versa.

Lastly, if $\delta y_i \rightarrow 0$ and $\delta y_j \rightarrow 0$, it could be inferred that some sort of quasi-equilibrium condition would be incurred.

Determination of Compositional Changes from the Material Balances

Knowing y_i, the behavior of x_i follows directly from the component material balances; that is, from the material balances, $F = V + L$ and $F(x_F)_i = Vy_i + Lx_i$, so that

$$\frac{V}{F} = \frac{(x_F)_i - x_i}{y_i - x_i}$$

or

$$x_i = -y_i \frac{V/F}{1 - V/F} + \frac{(x_F)_i}{1 - V/F}$$

where

$$x_i = (x_1)_i \quad \text{and} \quad y_i = (y_1)_i \text{ at point 1}$$

$$x_i = (x_2)_i \quad \text{and} \quad y_i = (y_2)_i \text{ at point 2}$$

Thus, each successive value of V/F can be used to obtain a new value of y_i and then of x_i. In turn, new values for Φ_i and Φ_j can be calculated, leading to the determination of the incremental area.

Determination of Area in Terms of V

As indicated previously, the more-ready determination for area utilizes the numerical integration of the expression

$$\frac{1}{\Phi_i + \Phi_j} \delta V = \delta A$$

The behavior of Φ_i and Φ_j ultimately are known in terms of V or V/F, through the medium of y_i and x_i.

Alternately, knowing the behavior of δy_i with respect to $\delta(V/F)$ or δV, it follows that the corresponding values of A can be determined from the numerical integration of

$$\delta A = F \frac{\dfrac{V}{F}\dfrac{\delta y_i}{\delta(V/F)} + y_i}{\Phi_i} \delta(V/F)$$

This is a less-direct, more-involved route and more subject to cumulative error.

Bidirectional Transfer

In the event a permeate phase V_1 is introduced, this may or may not induce a transfer of one or the other of the components in the opposite direction, depending on the mole-fraction composition. Whether or not such a reverse transfer can occur, of course, depends on the permeability characteristics of the membrane material.

For the record, a comparison of the conditions for concurrent flow at point 1 is as follows for the transfer of components i and j:

$$[P_L(x_F)_i \text{ minus } P_V(y_1)_i] \qquad \text{where } (x_F)_i = (x_1)_i$$

$$[P_L(x_F)_j \text{ minus } P_V(y_1)_j] \qquad \text{where } (x_F)_j = (x_1)_j$$

Whenever this difference is positive, the expected transfer of the component to the permeate phase occurs and the differential $d(Vy)$ increases. If negative, then transfer of the component from the permeate phase to the reject phase occurs and $d(Vy)$ decreases. Furthermore, presumably, one component may transfer in one direction, the other component in the opposite direction. The fundamental differential rate equations in each case remain consistent, and the differential dA remains positive in the conventions used.

Note, moreover, that both $d(Vy_i)$ and $d(Vy_j)$ can be positive at the same time or both negative at the same time. This cannot be the case for the mole fractions per se, however, for if, say, y_i increases, then y_j must decrease and vice versa.

The behavior of V may be judged from the equality that

$$d(Vy_i) + d(Vy_j) = dV$$

that is, depending on the sign and magnitude of the differentials on the left, the permeate rate V may increase or decrease.

It may be added that, if the difference approaches zero for one component, then consistency requires that the difference approach zero for the other component. As indicated elsewhere, this can be construed as a condition of quasi-equilibrium and the membrane areal requirement would increase without limit.

The phenomenon of reverse behavior for one component or the other is the principal feature of using recycle or reflux to achieve a sharper separation in multistage or cascade operations, as set forth in Chapter 4.

Some Further Comments, Particularly about Using V or V/F

Overall, the numerical integration can be viewed in terms of V or V/F, or in terms of $\ln V$ or $\ln V/F$, as the independent variable.

Note that the point where $\Phi_i = 0$ or $\Phi_j = 0$, by definition, is excluded from the determination, since infinite behavior would occur in one or another of the rate equations. Therefore, Φ_i and Φ_j cannot in general crossover and take on values of an opposite sign.

However, there can be a waxing or waning in, say, the behavior of y_i. In considerable part, this depends on the rate and inlet composition of the permeate phase V or V_1 relative to the feedstream F or L_1. Therefore, y_i could increase or decrease along with increasing or decreasing values of V, as could x_j, but not necessarily.

When $V_1 = 0$, and for total transfer where $L_2 = 0$, the final value of $y_i = (y_2)_i$ must be the same as that for the feedstream composition $(x_F)_i$. If a permeate phase V_1 is introduced, then there is an effective combined overall feed composition, as calculated from a combination of the feedstream F and the introduced permeate phase V_1. For total transfer, $y_i = (y_2)_i$ takes on this effective combined composition. As noted elsewhere, for most purposes, the calculation can be based on a unit feedstream molar flow rate; that is, $F = 1$.

If $V_1 = 0$, which is the usual case, then its composition can be determined from a bubble-point type calculation on $F = L_1$. The overall solution does not, in principle, invoke trial and error, save for the special case of establishing the initial bubble-point determination to determine $(y_1)_I$, starting at point 1 and proceeding to point 2.

Unfortunately, since the logarithmic behavior of V (or $1/V$) is involved in initiating and proceeding with the foregoing outline of the calculation, the point where $V_1 = 0$ is precluded as a starting point. The option therefore is to express the changes in terms of the reject rate L and its composition.

Alternately, of course, the integration could proceed backward from point 2, utilizing V as the variable. For complete transfer, V_2/F is unity, with the composition of L_2 determined from a dew-point type calculation on $F = V_2$; otherwise, this composition must be specified. However, V_1/F then becomes undefined.

Calculation Procedures Based on L or L/F

The foregoing equations may be rephrased in terms of L and x_i:

$$d(Vy_i) = -d(Lx_i) \qquad \text{and} \qquad d(Vy_j) = -d(Lx_j)$$

so that

$$\Phi_j \frac{L \, dx_i}{dL} + \Phi_j x_i = \Phi_i \left(-\frac{L \, dx_i}{dL} \right) + \Phi_i (1 - x_i)$$

or

$$L \frac{dx_i}{dL} (\Phi_i + \Phi_j) = -x_i (\Phi_i + \Phi_j) + \Phi_i$$

The signs for the terms in the latter equation(s) can sometimes be made more convenient by multiplying through by a negative one (-1). It turns out, however, on introducing F, that

$$\frac{\delta x_i}{\delta (L/F)} = \frac{F}{L} \left[-x_i + \frac{\Phi_i}{\Phi_i + \Phi_j} \right]$$

where $\delta x_j = -\delta x_i$. Here, the initial value of $L/F = L_1/F$ can be assigned unity and the initial boundary condition causes no problem in integrating away from point 1. The problem can occur at the other end, however, since a value of $L_2/F = 0$ is precluded; that is, complete transfer is not allowable.

From the material balances, $F = V + L$ and $F(x_F)_i = Vy_i + Lx_i$, so that

$$\frac{L}{F} = \frac{y_i - (x_F)_i}{y_i - x_i}$$

where

$$y_i = -x_i \frac{L/F}{1 - L/F} + \frac{(x_F)_i}{1 - L/F}$$

where

$$x_i = (x_1)_i \qquad \text{and} \qquad y_i = (y_1)_i \text{ at point 1}$$
$$x_i = (x_2)_i \qquad \text{and} \qquad y_i = (y_2)_i \text{ at point 2}$$

Thus, each successive value of L or L/F can be used to obtain a new value for x_i and then y_i. Last, we note that $dV = -dL$ so that

$$\frac{1}{\Phi_i + \Phi_j}(-dL) = dA$$

This is the most ready form for estimating successive changes in the area A.

The latter approach, using L instead of V, is utilized in the spreadsheet calculations of Appendix 6. Note that the initial composition for $L = L_1 = F$ is the feed composition, and a bubble-point type determination establishes the initial composition $(y_1)_i$ of $V = V_1$. (And note that the bubble-point type determination assumes that $V''/F = 0$, where the notation V'' is used for the permeate flux in a flash-vaporization type calculation.)

Countercurrent Flow

In countercurrent flow, the rate equations read

$$\frac{-d(Vy_i)}{P_i(P_L x_i - P_V y_i)} = \frac{-d(Vy_j)}{P_j(P_L x_j - P_V y_j)} = dA$$

where integration is from point 1 to point 2, with A positive in the direction of integration, and Vy_i (and Vy_j) decrease as A increases in the direction of integration, whenever the denominator is positive. If the denominator should become negative, as after a crossover, then the opposite effect would exist. However, the point at which the denominator would become zero is not allowed. The inference, therefore, is that both components transfer from the reject phase or stream L to the permeate phase or stream V but at different rates. Alternately, integration may proceed from point 2 to point 1, where A would take on negative values in the direction of integration.

In either case, the relationship or derivative $\delta y_i/\delta(V/F)$ is exactly the same as for concurrent flow.

The values for the boundary conditions for V_1 and $(y_1)_i$ at point 1 are not known, however. Nor, equivalently, are the values for L_2 and $(x_2)_i$ at point 2. The situation becomes double trial and error for a two-component system. For three or more components, the difficulties are compounded.

Note, for instance, that, if $V_2 = 0$, then its composition introduces the extra element of trial and error, since its composition has to be determined from a bubble-point type calculation on stream L_2, but whose flow rate and composition are not yet known and must be assumed.

The overall material balances may be phrased specifically as

$$V_1 + L_2 = V_2 + L_1$$

$$V_1(y_1)_i + L_2(x_2)_i = V_2(y_2)_i + L_1(x_1)_i$$

or, more generally, as the differences

$$\Delta = L_1 - V_1 = L_2 - V_2 = L - V$$

$$\Delta(x_\Delta)_i = L_1(x_1)_i - V_1(y_1)_i = L_2(x_2)_i - V_2(y_2)_i = Lx_i - Vy_i$$

where $(x_\Delta)_i$ denotes the composition of the hypothetical difference point or difference quantity designated Δ. Accordingly, at any point on the transfer surface,

$$\frac{V}{\Delta} = \frac{x_i - (x_\Delta)_i}{y_i - x_i}$$

or

$$y_i = x_i \frac{V/\Delta + 1}{V/\Delta} - \frac{(x_\Delta)_i}{V/\Delta}$$

or

$$x_i = y_i \frac{V/\Delta}{V/\Delta + 1} + \frac{(x_\Delta)_i}{V/\Delta + 1}$$

As in concurrent flow, it can be derived also, for two components i and j, that

$$\frac{\delta y_i}{\delta(V/F)} = \frac{V}{F}\left[-y_i + \frac{\Phi_i}{\Phi_i + \Phi_j}\right]$$

or

$$\frac{\delta y_i}{\delta \ln(V/F)} = -y_i + \frac{\Phi_i}{\Phi_i + \Phi_j}$$

where $F = L_1$. Furthermore, V is again designated positive but increases from point 2 to point 1 and decreases from point 1 to point 2.

Numerical integration may proceed from point 1 to point 2 or vice versa. From 1 to 2, however, the membrane area A takes on positive values, as defined previously for concurrent flow. The solution is double trial and error, since, say, V_1 and $(y_1)_i$ must be assumed, or else L_2 and $(x_2)_i$.

If $V_2 = 0$, then its composition may be determined, for each trial, from a bubble-point type calculation on L_2, assuming the composition $(x_2)_i$. In this circumstance, it is preferable to integrate from point 2 to point 1.

If both V_2 and $(y_2)_i$ are specified—that is, an external permeate phase is introduced at point 2—the situation nevertheless remains a double trial-and-error calculation, because it is still necessary to assume V_1 and $(y_1)_i$, or else L_2 and $(x_2)_i$.

6.3 RECYCLE

Part (or conceivably all, in the limit) of either or both the permeate and reject products may be recycled or refluxed. For instance, in the case of concurrent flow, part of the reject product L_2 can be recycled to the feed F, where the composition and rate of L_1 would be affected. In effect the reject is making another pass through the membrane cell. For that matter, part of the permeate product V_2 can be recycled to constitute V_1. The beneficial effects are dubious and can be ascertained only by a more complex mode of calculation. For the case of assuming perfect mixing, the benefits are nil, as already discussed in Section 3.5 of Chapter 3.

In the case of countercurrent flow, part of the reject product L_2 can be recycled to the feed or part of the reject can be recycled to become V_2, the inlet permeate phase composition. Also in countercurrent flow, part of the permeate product V_1 can be recycled to the feed.

These options are diagrammed in Figure 1.6 and further discussed in Chapter 7. The results are, in the main, perceived as beneficial in enhancing the degree of separation.

Moreover, as will be shown in Chapter 7, the aforementioned juxtapositions may be combined to yield a sharper separation between components. The layout is similar in principle to a distillation column.

The exact calculations, needless to say, become increasingly complex and introduce the specter of multiple trial-and-error procedures. For this reason, simplifications are in order, which are presented and discussed subsequently. In another way of looking at it, these already have been introduced via the multistage or cascade representation provided in Chapter 4.

6.4 LIMITING CONDITIONS

The complexity of the calculations, particularly for countercurrent flow, make it advisable to consider the limiting situations, as follows, referring to Figure 6.1. The two diametric circumstances or scenarios are (1) only reject outflow is produced and (2) only permeate outflow is produced.

It is assumed that no permeate phase is introduced, whereby in concurrent flow $V_1 = 0$ and in countercurrent flow $V_2 = 0$.

However, for circumstance 1, the permeate outflow, albeit nil, has a composition. And, for circumstance 2, the reject outflow, albeit nil, has a composition. These respective compositions represent the maximum degree of separation for the feedstream mixture.

Thus, the limiting conditions presumably mark the limits for the degree of separation attainable, although the recovery in either the reject outflow or permeate outflow is nil. That is, the feed is regarded as recovered either as reject only or permeate only. In the former circumstance, the composition of the minute amount or "drop" of permeate is determined from a bubble-point type calculation on the feed. In the latter, everything is recovered as permeate, so that the last drop of reject transformed to permeate is determined by a dew-point calculation on the permeate, which necessarily is of the same composition as the feed. These are the two extremes. In the one case, the bubble-point type calculation applies, in the other the dew-point type calculation applies.

Bubble-Point vs. Dew-Point Type Calculations

For the two-component system i and j, it may be written that

$$y_i/x_i = K_i x_j/y_j = 1/K_j$$

where i is regarded the more-permeable component. These two expressions, therefore, give a spread indicating the degree of separation that presumably can be obtained for components i and j without the use of recycle or reflux. In the one case, the bubble-point type determination applies; in the other case, the dew-point type determination. Each gives different values for K_i and K_j.

The bubble-point and dew-point type calculations are presented in Chapter 3 for a single-stage separation with perfect mixing. For the bubble-point type determination, the criterion is $V/F = 0$, and it is possible to determine the flux V'' where

$$K_i = b/(V'' + a) K_j = d/(V'' + c)$$

where a, b, c, and d are quantities to be affixed. For the dew-point type calculation, the criterion is $V/F = 1$, and a procedure is also provided for determining a value for V''.

From Example 3.1, note that K_i has the greater value for $V/F = 1$, which is the criterion for the dew-point type calculation. Whereas, $1/K_j$ has the greater value for $V/F = 0$, which is the criterion for the bubble-point type calculation. The separations and recoveries are also set forth in the example.

Concurrent Flow

For the first circumstance, where no permeate outflow is produced, $V_2 \to 0$ and $F = L_1 = L_2$, and the bubble-point type determination on the composition of F determines the composition of V_2 (even though $V_2 \to 0$). The bubble-point calculation corresponds to $V/F = 0$, as derived and used previously. The reject and permeate compositions expectedly remain constant along the membrane axis of flow.

For the second circumstance, where no reject outflow is produced, $L_2 \to 0$ and $F = L_1 = V_2$, and the dew-point type calculation on the composition of V_2 determines the composition of L_2 (even though $L_2 \to 0$). Note that the compositions $(x_F)_i = (x_1)_i = (y_2)_i$ become equal. The dew-point type calculation corresponds to $V/F = 1$, as derived and used previously. The reject and permeate compositions vary along the membrane axis of flow.

Countercurrent Flow

For the first circumstance, where only reject outflow is produced, $F = L_1 = L_2$ and the degree of separation is obtained by a bubble-point calculation on $F = L_1$; that is, this determines the composition of the "drop" of permeate representing V_1 (even though $V_1 \to 0$). The result necessarily is the same as for concurrent flow. The reject and permeate compositions remain constant along the axis of countercurrent flow.

For the second circumstance, where only permeate outflow is produced, since $V_2 = 0$ and $L_2 \to 0$, it follows that $L_1 \to V_1$. It is as if the flow of the feedstream proceeds directly from V_1 to L_1. However, the compositions of the reject stream L and permeate stream V expectedly change along the linear axis of flow. We are therefore interested in how each stream composition may vary and the compositions at point 2. In effect, point 2 is a closed end, but there may be a composition gradient for each stream between 1 and 2 that, in a way, can also be construed as producing a separation.

The determination proceeds as follows. Since

$$L_1 - V_1 = L - V$$
$$L_1(x_1)_i - V_1(y_1)_i = Lx_i - Vy_i$$

then, at any point, $V = L$ and $y_i = x_i$ (also $y_i = x_i$ for a two-component system). The question is, what values of y_i and x_i are attained at point 2? Differentiating the preceding,

$$dV = dL$$
$$V\,dy_i + y_i\,dV = L\,dx_i + x_i\,dL$$

which, since in this case $L_1 = V_1$ and $L = V$, fortuitously combine to yield

$$V \, d(y_i - x_i) = -(y_i - x_i)dV$$

Integrating between limits and rearranging,

$$V(y_i - x_i) = V_1[(y_1)_i - (x_1)_i]$$

Since $(x_1)_i = (y_1)_i$, then at any point along the axis of flow, it would be required that

$$y_i = x_i$$

which is arbitrary, and the values presumably vary with position along the axis. This also infers that $K_i = 1$ and $K_j = 1$, which, in turn, requires that $P_i = P_j$, which causes a contradiction. It can be concluded that this limiting condition for countercurrent flow has no singular answer for the separation that could be attained.

There is an interesting conjecture, however. If K_i and K_j tend to behave as constants, the relative volatility form can be invoked:

$$\frac{y_i/x_i}{y_j/x_j} = \frac{K_i}{K_j} = \alpha_{i-j}$$

Substituting for x_i and x_j, the following is obtained:

$$y_i = \frac{\alpha_{i-j}}{1 - x_i + \alpha_{i-j}}$$

When $x_i = 1$, then $y_i = 1$; or else when $x_i = 0$, then $y_i = 0$. There is the hint that one of the pure components may emerge in the far reaches of the membrane cell. Assumably, this would be the component with the lower permeability, that is, component j. The extent of this effect no doubt depends in part on the degree of any forward and back mixing.

(In passing, it may be observed that, for the relative volatility type of calculation, the L/F, V/F, and L/V ratios in terms of, say, x_i can be obtained for any given or specified feed composition $(x_F)_i$ by utilizing the calculated compositional differences $y_i - x_i$, $y_i - (x_F)_i$, and $(x_F)_i - x_i$ as obtained from a rearrangement of the component material balance.)

In confirmation of the preceding, consider the countercurrent relationship

$$\frac{-d(Vy_i)}{P_i(P_L x_i - P_V y_i)} = \frac{-d(Vy_j)}{P_j(P_L x_j - P_V y_j)} = dA$$

If $x_i = y_i$ and $x_j = y_j$, then

$$\frac{d(Vy_i)}{P_i(P_L - P_V)} = \frac{d(Vy_j)}{P_j(P_L - P_V)}$$

or

$$\frac{V\,dy_i + y_i\,dV}{P_i y_i} = \frac{-V\,dy_i + (1 - y_i)\,dV}{P_j(1 - y_i)}$$

Rearranging, collecting terms, and introducing F,

$$\frac{P_j - (P_i - P_j)y_i}{(P_i - P_j)y_i(1 - y_i)}\,dy_i = \frac{d(V/F)}{V/F}$$

Integrating term by term, between the limits $(y_1)_i = (x_F)_i$ and y_i, gives

$$\frac{P_j}{P_i - P_j}\left[\ln\frac{y_i}{1 - y_i} + \ln(1 - y_i)\right]\Bigg|_{(y_1)_i = (x_F)_i}^{y_i}$$

It follows that

$$\ln\left[\frac{y_i}{1 - y_i}\frac{1 - (x_F)_i}{(x_F)_i}\right] + \ln\frac{1 - y_i}{1 - (x_F)_i} = \frac{P_i - P_j}{P_j}\ln(V/F)$$

or

$$\frac{y_i}{1 - y_i}\frac{1 - (x_F)_i}{(x_F)_i}\frac{1 - y_i}{1 - (x_F)_i} = (V/F)^{\frac{P_i - P_j}{P_j}}$$

where

$$\frac{y_i}{(x_F)_i} = (V/F)^{\frac{P_i - P_j}{P_j}}$$

Therefore, as $V/F \to 0$,

$$y_i = x_i = 0 \quad \text{and} \quad y_j = x_j = 1$$

That is, in countercurrent flow, where L_2 and V_2 equal zero, there could exist a point in the far reaches or "dead space" of the membrane cell where, theoretically at least, the pure component j exists. A membrane could, in this way, serve as a concentrator for the less-permeable component. If nothing else, it is an interesting speculation.

The membrane area requirement is of interest:

$$\frac{-d(Vy_i)}{P_i y_i (P_L - P_V)} = dA$$

It has been determined that, on solving for V and multiplying by y_i,

$$Vy_i = F[1/(x_F)_i]^{\frac{P_j}{P_i - P_j}}[y_i]^{\frac{P_i}{P_i - P_j}+1}$$

Differentiating,

$$d(Vy_i) = F\left(\frac{P_j}{P_i - P_j} + 1\right)[1/(x_F)_i]^{\frac{P_j}{P_i - P_j}}[y_i]^{\frac{P_j}{P_i - P_j}-1}$$

Substituting, the area is related by

$$-F\frac{\left(\dfrac{P_j}{P_i - P_j} + 1\right)[1/x_F)_i]^{\frac{P_j}{P_i - P_j}}[y_i]^{\frac{P_j}{P_i - P_j}-1}}{\dfrac{P_j}{P_i - P_j}P_i(P_L - P_V)}\, dy_i = dA$$

Integrating between limits,

$$-F\frac{\left(\dfrac{P_j}{P_i - P_j} + 1\right)[1/x_F)_i]^{\frac{P_j}{P_i - P_j}}[y_i]^{\frac{P_j}{P_i - P_j}-1}}{\dfrac{P_j}{P_i - P_j}P_i(P_L - P_V)}\Bigg|_{(y_1)_i=(x_F)_i}^{y_i} = A$$

Substituting the limits and simplifying,

$$F\frac{1-[y_i/(x_F)_i]^{\frac{P_j}{P_i - P_j}}}{P_i(P_L - P_V)} = A$$

If $y_i = (y_2)_i = 0$,

$$F\frac{1}{P_i(P_L - P_V)} = A$$

This establishes the limiting relationships for $V_2 = 0$ and $L_2 = 0$ in countercurrent flow, where $y_i = x_i$ and $y_j = x_j$. Conceivably, in the theoretical limit, in the far reaches or dead space of the membrane cell, $y_j = x_j = 1$.

It affords the interesting conjecture that, if the less-permeable component j is the objective of recovery, then a single membrane cell of sufficient linear dimensions could suffice if operated in countercurrent flow. A notable example is the methane-nitrogen separation, where the known membrane materials are more permeable to methane than nitrogen and have low selectivity. Thus, subquality natural gas is upgraded by removing the nitrogen as reject. Unfortunately, the bulk of methane-rich gas has to pass through the membrane. At that, however, it might be simpler than multistage or cascade operations.

Stripping vs. Rectification

Inasmuch as the less-permeable component j may conceivably exist in a pure form in the far reaches of the membrane cell, the preceding operation can be referred to as *stripping*. If the feed is introduced into the low-pressure side of the cell, and at the far end, a compressor is used to maintain the pressure on the high-pressure side, then the more-volatile component could conceivably tend to accumulate in the far reaches or "dead space" of the cell. This may be referred to as *rectification*.

The low-pressure side can be referred to as the *permeate side*, as before, and the high-pressure side as the *reject side*. In this case, the permeate phase, in effect, becomes the continuation of the feed. In the limit, there is no net production of permeate, and the final reject stream is of the same composition as the feed.

The derivations are fully analogous to those for stripping, and the areal requirement, in the limit, reduces to a relationship similar to stripping:

$$F \frac{1}{P_i (P_L - P_V)} = A$$

The membrane areal requirements are the same, but the more-permeable component could conceivably be obtained in the pure form.

It should be emphasized that this may be construed as a case where the bulk of the gas stream may not have to pass through the membrane proper. As such, it may be more applicable, say, to the separation of methane-rich gas from nitrogen. There is no doubt a trade-off.

The subject in its practical application becomes that of countercurrent flow with recycle, which is presented in the next chapter.

6.5 EQUILIBRIUM

As previously discussed, the requirement for equilibrium should be that the rate of mass transfer should be zero; that is, for components i and j,

$$\Phi_i = P_i(P_V y_i - P_L x_i) = 0$$

$$\Phi_j = P_j(P_V y_j - P_L x_j) = 0$$

where

$$\frac{y_i}{x_i} = \frac{P_L}{P_V} = \frac{y_j}{x_j} = \frac{1-y_i}{1-x_i}$$

A contradiction is produced, however, since on the one hand,

$$y_i - y_i x_i = x_i - y_i x_i \qquad \text{or} \qquad y_i = x_i$$

but on the other hand,

$$1 = y_i/x_i = P_L/P_V \neq 1$$

Quite obviously, then, for true equilibrium, it must be required that $P_V = P_L$; that is, the pressure must be uniform throughout the system, on both sides of the membrane. Furthermore, the composition is uniform throughout the system. Since, by definition, in membrane separations, $P_V \neq P_L$, a condition of true equilibrium cannot be reached.

EXAMPLE 6.1

The limiting information calculated and reported in Example 3.1 is repeated as follows. For bubble-point type calculation ($V/F = 0$),

$$K_i = 1.133536 \qquad 1/K_j = 1.097723$$

For dew-point type calculation ($V/F = 1$),

$$K_i = 1.142857 \qquad 1/K_j = 1.083333$$

The range for the degree of separation is poor under the conditions used, indicating that multistage or cascade operations are advisable for any sort of sharp separation.

For the recovery of the less-permeable component j as the reject in countercurrent flow, in the limit, there is the speculation that the degree of separation could approach $1.0/0.6 = 1.67$.

The corresponding spreadsheet-type calculations are shown in Appendix 6, including the determination for membrane area.

REFERENCES

1. Hoffman, E. J. *Heat Transfer Rate Analysis.* Tulsa, OK: PennWell, 1980.
2. Hoffman, E. J. *Azeotropic and Extractive Distillation.* New York: Wiley-Interscience, 1964; Huntington, NY: Krieger, 1977.

7

Countercurrent Flow with Recycle

The preceding chapter shows that, for the limiting conditions of countercurrent flow with no reject produced, the less-permeable component hypothetically could concentrate in the far end of the reject side of the cell. Similarly, with no net permeate produced and compression from the permeate side to the reject side, it is theoretically conceivable for the more-permeable component to concentrate in the far end of the permeate side of the cell.

In practice, these conditions could be accented by the use of recycle, in the first instance part (or all, in the limit) of the reject, and the operation may be regarded as "stripping." In the second instance, part (or all, in the limit) of the permeate on the low-pressure or permeate side is compressed and recycled to the high-pressure or reject side, and the operation may be regarded as "rectification." The two situations are diagrammed and compared in Figure 7.1.

Both stripping and rectification may be combined, as shown in Figure 7.2, to produce a sharper separation. The overall operation then corresponds to a distillation column, such as a packed or wetted-wall column, which can be described in terms of a continuum.[1] In distillation and in absorption and stripping, the calculation methodology is to assume constant internal flow rates, which in general does not apply here but can be utilized as a workable simplification.

Each section (the stripping section and the rectifying section) may be sized by the approximate formulas previously derived, that for rectification being

$$F \frac{1}{P_j(P_L - P_V)} = A$$

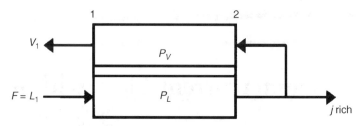

(a) Stripping in countercurrent flow

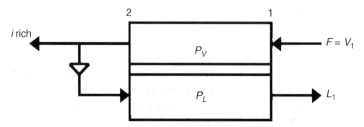

(b) Rectification in countercurrent flow

Figure 7.1 Use of recycle in countercurrent flow.

and for stripping,

$$F \frac{1}{P_i(P_L - P_V)} = \overline{A}$$

where P_j and P_i are the overall permeability coefficients. These results, in principle, refer to a nominal membrane area A and $\overline{A}$ for each section, assuming that the pure components could be attained, or nearly so.

7.1 CONSTANT FLOW RATES

As a footnote, considerable simplification in the mathematical sep-aration representations could result by assuming that the respective molar flow rates remain constant throughout the membrane unit. Such is the practice in distillation calculations, where there is mass transfer in both directions.[1] The assumption is similarly made in absorber or stripper calculations, where only one key component is involved.[1-4] This condition, called *constant molal overflow* in distillation and absorber and stripper derivations and calculations, may also be accommodated in the case of multistage or cascade membrane calculations, as derived and utilized

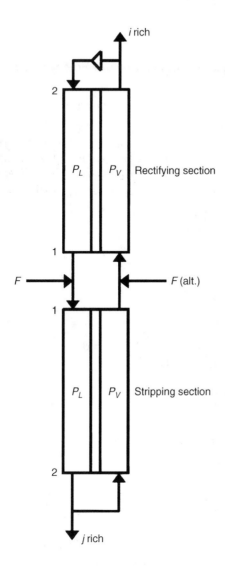

Figure 7.2 Combined stripping and rectification.

in Chapter 4. There, perfect mixing is assumed and the recycle rate between stages is adjustable due to the use, and necessity, of interstage compression to raise recycled permeate phase pressure back to the reject phase level for the next successive stage.

As will be demonstrated, the great advantage in assuming constant molal or molar flow rates is that analytic integration can be performed on the rate equations in lieu of the more comprehensive and rigorous numerical methods.

*Rate Equations for Constant Molar
or Molal Flow Rates*

The rate equations may be phrased similarly to those presented and utilized in Chapter 6. Here, however, the appropriate notation is used to distinguish the rectifying and stripping sections for an overall or combined operation with intermediate feed.

In the rectifying section, the more-permeable component i is assumed controlling and preferred for integration. Component j more or less is viewed as a dependent variable, obtained by difference. In the stripping section, the less-permeable component j is assumed controlling and preferred for integration. Component i more or less is viewed as a dependent variable, obtained by difference.

Rectifying Section

For the rectifying section, sometimes called the *absorbing section*, at constant V (and L),

$$\frac{V dy_i}{P_i(P_L x_i - P_V y_i)} = \frac{V dy_j}{P_j(P_L x_j - P_V y_j)} = dA$$

where $y_i + y_j = 1$, and $dy_i + dy_j = 0$. Alternately,

$$V \frac{dy_i}{dA} = P_i(P_L x_i - P_V y_i) = \Phi_i$$

$$V \frac{dy_j}{dA} = P_j(P_L x_j - P_V y_j) = \Phi_j$$

Integration may proceed up or down the column, but here dA is assumed positive in the upward direction; that is, the tubular membrane area A increases in the direction of integration. Thus, in the rectifying section, integration proceeds from the feed location "upward" toward the more-permeable product end, designated product D. (Note that this convention is opposite to that used in numbering the stages in the stagewise calculations of Chapter 4, where stage numbering starts at the top or more-permeable product end and proceeds "downward" toward the feed location.) For the purposes here, it therefore seems more natural to integrate starting from the feed location.

The convention used here is that y_i increases (and y_j decreases) as A increases, which corresponds to an integration of the membrane rectifying section "upward" from the central or intermediate feed location (point 1) toward the more-permeable product end (point 2). In the limit, for the more-permeable product, $(y_2)_i = (x_2)_i = (x_D)_i$.

Note, further, that, if Φ_i is positive, then Φ_j has to be negative. However, this is a contradiction brought about by the assumption of constant molar flow rates, because in theory, Φ_j is also positive. The more-permeable component i is merely transferred to the permeate faster than component j, resulting in an increasing concentration of the former.

For obvious reasons, therefore, the second equation involving component j must be ignored, with the first equation assumed controlling. The net effect in proceeding "upward" is that component i is absorbed from the reject phase into the permeate phase, whereas component j merely appears to be stripped from the permeate phase and absorbed into the reject phase, even though the concentration of j in the reject increases with the downward flow of the reject phase.

Feedstream Composition and Feed Location

Moreover, by analogy with the stagewise calculations of Chapter 4, for simplicity, it is assumed that the feedstream is introduced on the reject side. Moreover, the reject stream compositions are regarded as being known at the feed location, for both the rectifying and stripping sections, and identical with the feedstream composition. This is entirely analogous to the assumptions used for stagewise calculations in Chapter 4.

In turn, the permeate phase compositions for both the rectifying and the stripping sections are assumed to have the same common value at the feed location; here, the requirement for continuity. This permeate composition can be determined by a flash-type calculation on the feedstream composition, in particular, a bubble-point type calculation.

At the same time, the K-values are established, as is the molar flux V''. Not only this, but by assuming that the feedstream is introduced totally on the reject side, the flux V'' retains the same value in both the rectifying and stripping sections. Again, the analogy is with the stagewise calculations of Chapter 4.

Stripping Section

Likewise, in the overbar or overline notation for the stripping section, it can be written that

$$\frac{\bar{L}\,d\bar{x}_i}{P_i(P_L\bar{x}_i - P_V\bar{y}_i)} = \frac{\bar{L}\,d\bar{x}_j}{P_j(P_L\bar{x}_j - P_V\bar{y}_j)} = d\bar{A}$$

where $\bar{x}_i + \bar{x}_j = 1$ and $d\bar{x}_i + d\bar{x}_j = 0$. Alternately,

$$\bar{L}\frac{d\bar{x}_i}{dA} = P_i(P_L\bar{x}_i - P_V\bar{y}_i) = \overline{\Phi}_i$$

$$\bar{L}\frac{d\bar{x}_j}{dA} = P_j(P_L\bar{x}_j - P_V\bar{y}_j) = \overline{\Phi}_j$$

Here, the convention is used that $\bar{x}_j$ increases (and $\bar{x}_i$ decreases) as $\bar{A}$ increases, which is used to connote an integration of the membrane stripping section "downward" from the central feed location (point 1) toward the less-permeable bottoms product end (point 2). In the limit, of course, $(\bar{x}_2)_j = (x_B)_j = 1$. The convention used in the stripping section, therefore, is that $\bar{A}$ is positive and increasing toward the less-permeable product end, designated as the bottom.

Observe that $\overline{\Phi}_i$ is positive and $d\bar{x}_i$ is negative, which produces a contradiction, whereas $\overline{\Phi}_j$ is also positive and so is $d\bar{x}_j$. For obvious reasons, the first equation involving component i must be ignored, with the second equation controlling.

That is, in proceeding "downward," component i is stripped or deabsorbed faster from the reject phase than component j. The overall effect is that component j appears concentrated in the reject phase. In the opposite direction, however, in the "up-flowing" permeate phase, component i tends to be concentrated, whereas component j appears stripped.

As noted elsewhere, the foregoing assumed directional juxtapositions for integration are the inverse of those used in enumerating the stages in the rectifying and stripping sections of a multistage operation, as per Chapter 4.

Material Balances

In general, for countercurrent flow, at constant V and L, it may be written for the rectifying section, most simply, that

$$Vy_i = Lx_i + Dx_D$$

from which

$$Vdy_i = Ldx_i$$

Both the lower limits $(y_1)_i$ and $(x_1)_i$ correspond to the permeate and reject compositions from a bubble-point type calculation of the feedstream at the feed location. For the upper limits or values,

$$(y_2)_i = (x_2)_i = (x_D)_i$$

when external recycle or reflux is used.

Solving for x_i from the algebraic or integrated expression,

$$x_i = \frac{V}{L}y_i - \frac{D}{L}(x_D)_i$$

Note that, at the upper limit or limits, $V = L + D$.

In the stripping section, for component j,

$$\bar{L}\bar{x}_j = \bar{V}\bar{y}_j + B(x_B)_j$$

or

$$\bar{y}_j = \frac{\bar{L}}{\bar{V}}\bar{x}_j - \frac{B}{\bar{V}}(x_B)_j$$

where the range of variation is from the central feed location toward the less-permeable bottoms product end. Moreover, for partial recycle of the less-permeable bottoms product B,

$$(\bar{x}_2)_j = (\bar{x}_2)_j = (x_B)_j$$

that is, the mole fractions of the bottoms product streams are equal. Note that, at this point, the overall material balance for the stripping section is obtained: $\bar{L} = \bar{V} + B$.

Constraints, Contradictions, and Inconsistencies

Interestingly, for the situation at hand, a further constraint is produced at constant V (and L) for the two key components. For the rectifying section, at constant V (and L), in the conventions used,

$$Vdy_i = P_i(P_L x_i - P_V y_i)dA = \Phi_i \, dA$$

$$Vdy_j = P_j(P_L x_j - P_V y_j)dA = \Phi_j \, dA$$

Since $dy_j + dy_i = 0$, and $Vdy_i = -Vdy_j$, it therefore follows that

$$-\Phi_i = \Phi_j$$

that is, when the transfer of component i is in one direction, then the transfer of component j presumably is in the opposite direction.

It may be further added, that since $Vdy_i = Ldx_i$, then $Vdy_i = -Ldx_j$ and vice versa.

This rate equality translates to

$$-P_i(P_L x_i - P_V y_i) = P_j(P_L x_j - P_V y_j)$$

$$= P_j[P_L(1 - x_i) - P_V(1 - y_i)]$$

Collecting terms,

$$-P_i(P_L x_i - P_V y_i) = P_j(P_L - P_V) - P_j(P_L x_i - P_V y_i)$$

or

$$-\Phi_i = P_j(P_L - P_V) - (P_j/P_i)\Phi_i$$

or

$$\Phi_i[-1 + (P_j/P_i)] = P_j(P_L - P_V)$$

or

$$\Phi_i = \frac{P_i P_j}{-P_i + P_j}(P_L - P_V)$$

In other words,

$$P_L x_i - P_V y_i = \frac{P_j}{-P_i + P_j}(P_L - P_V)$$

The foregoing infers that the difference $(P_L x_i - P_V y_i)$ necessarily has to remain constant. Interestingly, however, since $P_i > P_j$, the difference is negative, implying a transfer of component i from stream V to stream L, which is the reverse from what is normally expected, as per the rate equations; hence, a contradiction.

Moreover, it has already been shown from the rate relationships that $\Phi_j = -\Phi_i$, such that

$$P_L x_j - P_V y_j = -\frac{P_j}{P_i}(P_L x_i - P_V y_i)$$

where the difference $(P_L x_j - P_V y_j)$ also is constant and has a negative value, if the difference $(P_L x_i - P_V y_i)$ is positive. That is to say, if so, the transfer of component j is from the permeate stream V to the reject stream L. Keep in mind that these are the conditions at steady state, after the concentration or mole fraction profiles have been stabilized.

Still another way of regarding these relationships is to start with

$$\Phi_j = P_j(P_L x_j - P_V y_j) = P_j[P_L(1 - x_i) - P_V(1 - y_i)]$$
$$= P_j(P_L - P_V) - P_j(P_L x_i - P_V y_i)$$

where

$$\Phi_j = P_j(P_L - P_V) - (P_j/P_i)\Phi_i \quad \text{or} \quad 0 = P_j(P_L - P_V) - (P_j/P_i)\Phi_i - \Phi_j$$

where Φ_i and Φ_j are defined as before, in Chapter 6.

It appears that $\Phi_i \neq -\Phi_j$. There is an apparent contradiction. By dividing through by P_j, the preceding relationship transforms to

$$(P_L x_i - P_V y_i) + (P_L x_i - P_V y_i) = P_L - P_V$$

which, after summing the values of x and y, is nothing more than an identity. Nevertheless, the implication remains that, if component i, say, is transferred in one direction, component j transfers in the other and vice versa.

In conclusion, it would be required that $P_L x_i < P_V y_i$, so that transfer of the more-permeable component i occurs from the permeate to the reject phase. At the same time, the less-permeable component j must transfer from the reject phase to the permeate phase.

Alternately, on switching i and j, we obtain

$$0 = P_i(P_L - P_V) + (P_j - P_i)(P_L x_j - P_V y_j)$$

Here, since $P_i > P_j$, it follows that $P_L x_j > P_V y_j$, which is in agreement with the previous paragraph but not necessarily with the usual interpretation for the relative permeability behavior for the respective components.

For these contradictory reasons, the assumption of constant molal behavior is generally applied to one component only; that is, the rate and material balances are confined to one component only, viewed as the controlling or more-significant component of the mixture. (In the rectifying section, this is likely the more-permeable component i; in the stripping section, component j.) Therefore, the behavior of the other component is viewed as dependent. In other words, the mole fractions, say, of the other component are those left over at the end of the calculations. In another way of saying it, by assuming constant molal or molar behavior, we over-specify the system; that is, use more than the allotted degrees of freedom or add too many equations, such as $V = $ constant, $L = $ constant.

All said, compositions or mole fractions may be found so that the preconceived conditions are met. Therefore, if the permeate phase becomes richer in component i as it moves "up" in the rectifying section and the reject phase becomes richer in component j as it moves "down" in the stripping section, the stated conditions may be said to prevail or to provide a useful approximation.

This sort of richness and leanness can be demonstrated in cascade or stagewise operations, where both a permeate phase and reject phase are introduced or recycled (or refluxed) into successive membrane cells. In this respect, the use of constant molar or molal flow rates in Chapter 4 for stagewise operations may not be too far afield.

Lastly, in each section of the continuum, the compositions of the permeate and reject streams or phases may tend to offset one another. In other words, there is a "window of opportunity" where interphase transfer can be bidirectional for one or the other of the components.

7.2 ANALOGY WITH WETTED-WALL DISTILLATION

Here, the embodiment is analogous to a wetted-wall distillation column with external reflux and reboil.[1] The membrane surface in effect replaces the (hypothetical) film between the liquid and vapor phases, this film connotes the resistance or conductance to mass transfer. The membrane unit proper can be viewed as a vertical or horizontal cylinder, the former placement more closely resembling a distillation column. The reject phase L is on the outside of the membrane cylindrical surface, between the membrane and the outer enclosing wall; the permeate phase V is on the inside of the membrane cylindrical surface. This juxtaposition corresponds to the liquid phase L wetting the inside of the distillation column and flowing downward by gravity and the vapor phase V moving upward in the column interior. (The juxtaposition could be reversed in the case of membrane columns.) For purposes of simplicity, both V and L are perceived as remaining constant up and down the membrane column, the same simplification used for wetted-wall distillation.

The feedstream is introduced at an intermediate position, by definition between the rectifying section and stripping section. It is further assumed that the feedstream is introduced into the reject phase and the reject phase at this point has the same composition as the feedstream. Furthermore, the reject phase (i.e., feedstream) is at its "bubble-point" so that a bubble-point type calculation necessarily yields a stream (the reject phase) with the same composition as the feed and another stream (a permeate phase) with the composition obtained by the calculation.

The further consequence is that the flow rate of the reject phase in the stripping section is equal to the flow rate of the reject phase in the rectifying section plus the feedstream rate, whereas the flow rate of the permeate phase is the same in both sections.

The permeate composition so obtained is regarded equal to that of stream V at the feed location and designated $(y_1)_i$, where component i is the more-permeable component. This is the lower limit of integration for the rectifying section. The reject composition so obtained is regarded equal to that of stream $\overline{L}$ at the feed location and designated $(\overline{x}_1)_j$, where component j is the less-permeable component. This is the lower limit of integration for the stripping section. The relative stream rates are determined from the reflux or reflux ratio L/D, as spelled-out in Example 4.1 and in Appendix 4.

For the record, as previously enunciated, $\overline{V} = V$, and $\overline{L} = L + F$. Furthermore, $V = L + D$, so that

$$\frac{L}{V} = \frac{1}{1 + \dfrac{L}{D}}$$

where L/D (or L/V) is specified or assigned.

In turn, for the stripping section,

$$\frac{\overline{L}}{\overline{V}} = 1 + \frac{1}{\dfrac{\overline{V}}{B}} \quad \text{or} \quad \frac{\overline{V}}{\overline{L}} = \frac{1}{1 + \dfrac{1}{\overline{V}/B}}$$

where $\overline{V}/B$ (or $\overline{V}/\overline{L}$ or $\overline{L}/\overline{V}$) is specified or assigned.

Since $V'' = \overline{V}''$ is already known from a the bubble-point type determination on the feedstream, it follows that

$$L'' = (L/V)V''$$

$$\overline{L}'' = (\overline{L}/\overline{V})\overline{V}''$$

$$F'' = \overline{L}'' - L''$$

In turn,

$$D'' = L'' \frac{1}{L/D}$$

$$B'' = \overline{V}'' \frac{1}{\overline{V}/B}$$

Thus, all stream quantities can be calculated on a flux basis. As a check, $F'' = D'' + B''$. Furthermore, $D/F = D''/F''$ and $B/F = B''/F''$, and so forth, so that all stream flow rates can be referenced to the feedstream flow rate (which may be assigned an arbitrary value of unity).

On determining $V'' = \overline{V}''$ from the bubble-point type calculation on the feedstream F, the corresponding value of D can be determined from the previously assigned external reflux or recycle ratio L/D and the value of B from the assigned external recycle ratio $\overline{V}/B$. In turn, L and $\overline{L}$ can be determined, where as previously noted, $\overline{L} = L + F$ and $\overline{V} = V$. It is "built in" that the various flow rates are mutually consistent.

The next matter to be taken up is the degree of separation that can be attained, and so forth, as previously demonstrated in Example 4.1 for stagewise separations and in the spreadsheet calculations of Appendix 4. In fact, the forepart of Appendix 4 is utilized in the spreadsheet calculations of Appendix 7 for the membrane viewed as a continuum, with recycle or reflux at both ends.

7.3 INTEGRATION OF THE FUNDAMENTAL RATE EQUATIONS

The rate equations can be integrated analytically by substituting the corresponding material balance in place of a mole fraction term. The integration is performed for both the rectifying section and the stripping section. The resulting behavior is logarithmic or exponential.

Rectifying Section

On substituting the material balance for x_i and dropping the component subscript i for the mole fractions, the integrated rate equation for the rectifying section becomes

$$\int_1^2 \frac{V\,dy}{P_i\left\{P_L\left[\dfrac{V}{L}y - \dfrac{D}{L}x_D\right] - P_V y\right\}} = \int_0^{A2} dA$$

$$\frac{V}{P_i}\int_1^2 \frac{dy}{\left(P_L\dfrac{V}{L} - P_V\right)y + P_L\dfrac{D}{L}x_D} = A_2$$

Note here that dy for the more permeable component i is positive in the "upward" direction of integration and that denominator also must be positive; that is, the transfer of component i is from the reject phase to the permeate phase. (The inference is that the less-permeable component *j* behaves oppositely.) Performing the integration,

$$\frac{V}{P_i} \frac{1}{\left(P_L \dfrac{V}{L} - P_V\right)} \ln\left\{\left(P_L \frac{V}{L} - P_V\right) y + P_L \frac{D}{L} x_D\right\}\Bigg|_1^2 = A_2$$

This furnishes a determination (i.e., estimation) for the membrane area A_2 in the rectification section based on component *i*. Integration is from the feed location toward the more-permeable product end, designated *D*, with *A* viewed as positive.

Limitations for Rectifying Section
Observe that, if no transfer of component *i* takes place, then

$$\left(P_L \frac{V}{L} - P_V\right) y + P_L \frac{D}{L} x_D = 0$$

Since $V = D + L$ and $V/L = D/L + 1$, the processes of substitution, rearrangement, and collection of terms eventually yield

$$1 - \frac{P_V}{P_L} = \frac{D}{L}\left(-1 - \frac{x_D}{y}\right)$$

which is an obvious contradiction. A similar contradiction occurs for the stripping section. That this kind of contradiction can occur is indicative of the limitations of assuming constant molar flow rates and the fact that zero transfer is not accommodated by the theory or, at least, is not allowable.

Stripping Section
For the stripping section, the rate balance can be rewritten in terms of the less-permeable component *j*, as follows. On substituting the material balance for y_i and dropping the component subscript *j* for the mole fractions, the integrated rate equation for the stripping section becomes

$$\int_1^2 \frac{\overline{L}\, d\overline{x}}{P_j\left\{P_L \overline{x} - P_V\left[\dfrac{\overline{L}}{\overline{V}}\overline{x} - \dfrac{B}{\overline{V}}x_B\right]\right\}} = \overline{A}_2$$

or

$$\frac{\overline{L}}{P_j} \int_{1}^{2} \frac{d\overline{x}}{\left(P_L - P_V \frac{\overline{L}}{\overline{V}}\right)\overline{x} + P_V \frac{B}{\overline{V}} x_B} = \overline{A}_2$$

where $\overline{A}_2$ denotes the area. Note that $d\overline{x}$ for component j is positive in the "downward" direction of integration and the denominator must also remain positive; that is, component i transfers more rapidly from the reject to the permeate phase, resulting in an accumulation or buildup of component j in the reject phase. (Despite a transfer of component j as well to the permeate, the buildup of component i in the permeate is greater than component j. This characteristic, of course, compromises the assumption of constant molar flow rates.)

If integration were upward, in the same direction as for the rectifying section, then $d\overline{x}$ for component j would be negative and a negative sign would necessarily be introduced, if the area were to remain positive.

Performing the integration "as is,"

$$\frac{\overline{L}}{P_j} \frac{1}{\left(P_L - P_V \frac{\overline{L}}{\overline{V}}\right)} \ln\left\{\left(P_L - P_V \frac{\overline{L}}{\overline{V}}\right)\overline{x} + P_V \frac{B}{\overline{V}} x_B\right\}\Big|_{1}^{2} = \overline{A}_2$$

This furnishes a determination for the membrane area $\overline{A}_2$ in the stripping section based on component j. Integration is from the feed location "downward" toward the less-permeable product end, designated B, with $\overline{A}$ perceived as positive.

Limitations for the Stripping Section

If no transfer of component j occurs, then it is required that

$$\left(P_L - P_V \frac{\overline{L}}{\overline{V}}\right)\overline{x} + P_V \frac{B}{\overline{V}} x_B = 0$$

which creates a situation of indeterminacy. Continuing, however, since $\overline{L}/\overline{V} = 1 + B/\overline{V}$ and on chasing through the subsequent rearrangements, it follows that, for component j,

$$\frac{P_L}{P_V} - 1 = \frac{B}{\overline{V}}\left(1 - \frac{x_B}{\overline{x}}\right) \quad \text{or} \quad \frac{P_L}{P_V} - 1 = \frac{\left(1 - \frac{x_B}{\overline{x}}\right)}{\overline{V}/B}$$

which, given $\bar{x} = (x_F)_j$, places a contradiction on the assigned value for $(x_B)_j$, since it is required that $x_B > \bar{x}$ for component j.

EXAMPLE 7.1

The same membrane characteristics and operating conditions are used as in the case of Example 4.1, which are the same as for Example 3.1. In Example 4.1, however, the K-value concept was utilized to establish stagewise absorption and stripping factors for the rectifying and stripping sections, where for component i,

$$K_i = \frac{P_L P_i}{V'' + P_V P_i}$$

and $y_i = K_i x_i$, similarly for component j.

The concept of the gas film mass transfer coefficient is customarily employed in the continuum theory of absorption and stripping and distillation, most often designated as k_g or as $(k_g)_i$ for a component i. It is analogous to what is called the *film heat transfer coefficient*. In the case of a packed column, it may be based on the superficial volume of the column and designated simply $k_g a$ on dropping the component subscript. (In fact, only one key component is under consideration and in relatively minor concentrations, with the other components customarily regarded as inert.) As applied, it is assumed that the liquid phase present is at equilibrium with a gas film, with this equilibrium designated $y^* = mx$, where the correlating coefficient m is treated as a constant. It may be approximated by Henry's law in the equivalent and appropriate units, for instance, or even by Raoult's law, leading to the use of the K-value instead of the coefficient m. The driving force for absorption is then $y-y^*$, or in terms of partial pressure, $P(y-y^*)$. The mass or molar transfer rate is viewed as gas film controlling and proportional to the gas film mass transfer coefficient k_g. The subject is discussed at some length in Hoffman[1] and other references and is not further pursued here. Instead, application of the two-component membrane permeability concept is considered more appropriate, even though it has its own shortcomings.

Bubble-Point of Feed and Feed Location

It was determined in the bubble-point type calculation for Example 3.1 and reiterated in Example 4.1, where $V/F = 0$, that

$$V'' = 12.9317$$
$$K_i = 1.133536 \qquad K_j = 0.910977$$
$$1/K_i = 0.882195 \qquad 1/K_j = 1.097723$$

This information is utilized at the feed location, where it is assumed that the reject stream L has the same composition as the feedstream, $(x_F)_i$ and $(x_F)_j$, and the permeate stream V takes on the bubble-point values $y_i = K_i x_i$ and $y_j = K_j x_j$. These values represent the limits for integration from the feed location, with integration proceeding "outward" from the feed location, as the case may be. That is, we can speak of the limiting value for the rectifying section at the feed location as $(y_1)_j$, with the upper limit at point D being $(y_2)_j$, similarly for the stripping section, where the limiting value at the feed location can be designated $(\bar{x}_1)_j$, with the "upper" limit at point B being $(\bar{x}_2)_j$.

As a matter of reference, there could be an optimum feed location, the location that gives the minimum column size for a given separation. This condition is ordinarily assumed to exist when both the rectifying section composition(s) and stripping section composition(s) at the feed location are approximately the same composition as the feedstream. In multicomponent distillation calculations, for instance, this gives rise to the necessity of the rectifying section calculation meshing to some degree with the stripping section calculation, in particular for the key components.

Another feature of distillation calculations, as ordinarily scoped, is to start from the top or distillate product end and work downward toward the feed location and start from the bottoms product end and work upward toward the feed location. In plate-to-plate or stage-to-stage calculations, this brings up the matter of meshing and an integral numbers of steps or stages.[1]

For the purposes here, however, the membrane determinations start from the feed location and work "outward"; that is, "upward" in the rectifying section and "downward" in the stripping section. This orientation is used in Example 4.1. An advantage is that "meshing" is already assured, even though it utilizes the feedstream composition and its bubble point. (The analogy is with the McCabe-Thiele method for binary distillation calculations, of Chapter 4, where it may be assumed that the feed is a saturated liquid.) It then becomes a matter of relating the degree of separation and the number of stages (as in Example 4.1) or membrane area (as in this example).

It may be added that, whereas in Example 4.1 the number of stages is assumed and the separation then calculated, here the separation is assumed and the total membrane area for each section then calculated. This route is made necessary on account of the logarithmic nature of the analytically integrated solutions.

Beyond all this is the notion that the enthalpy balances should be included as well as the material balances, which of course, can make things really complicated, assuming in the first place that the enthalpy behavior is reliably known as a function of composition. In binary distillation calculations, this generalization is embodied in the so-called Ponchon-Savarit method, as distinguished from the McCabe-Thiele method, which employs only material balances.[1,2] Both of these methods are discussed in Chapter 4.

Further derivations and the corresponding spreadsheet calculations for Example 7.1 are presented in Appendix 7. The membrane properties are made the same as for Example 4.1, and the reflux ratio L/D is first assigned. The solution ultimately becomes trial and error in the product streams.

Feed Composition and Stream Rates as per Example 4.1

The feed composition is specified as

$$(x_F)_i = 0.4 \quad \text{and} \quad (x_F)_j = 0.6$$

Assuming $L/V = 0.4$ for the first trial, the absorption factors are:

$$A_i = (0.4)(0.882195) = 0.352878$$

$$A_j = (0.6)(1.097723) = 0.439089$$

and it follows that

$$1 - (A_i)^6 = 0.9980692 \qquad 1 - (A_j)^6 = 0.9928333$$
$$[1 - (A_i)^5](0.4) = \underline{0.3978113} \quad [1 - (A_j)^5](0.4) = \underline{0.3934714}$$
$$\text{Difference} \quad 0.6002579 \qquad\qquad\qquad 0.5993619$$

Accordingly,

$$(x_D)_i = K_i(x_F)_i \frac{1 - A_i}{(\text{Difference})_i} = (1.133536)(0.4)\frac{1 - 0.352878}{0.6002579} = 0.488814$$

$$(x_D)_j = K_i(x_F)_j \frac{1 - A_j}{(\text{Difference})_j} = (0.910977)(0.6)\frac{1 - 0.439089}{0.5993619} = 0.511521$$

For the sum on the right,

$$\sum = 1.000335$$

This is regarded as sufficiently close for the purposes here. Note that, for a given value of n, this sum increases as L/V is increased.

For the stripping section, also assuming that here $\overline{V}/\overline{L} = 0.4$ for the first trial, the stripping factors are:

$$S_i = 0.4(1.133536) = 0.453414$$

$$S_j = 0.4(0.910977) = 0.364391$$

and it follows that

$$1 - (S_i)^6 = 0.991311 \qquad\qquad 1 - (S_j)^6 = 0.997659$$
$$[1 - (S_i)^5](0.4) = \underline{0.392335} \qquad [1 - (S_j)^5](0.4) = \underline{0.397430}$$
$$\text{Difference} \quad 0.598976 \qquad\qquad\qquad\qquad 0.600229$$

Accordingly,

$$(x_B)_i = (x_F)_i \frac{1 - S_i}{(\text{Difference})_i} = 0.4\frac{1 - 0.453414}{0.598976} = 0.365014$$

$$(x_B)_j = (x_F)_j \frac{1 - S_j}{(\text{Difference})_j} = 0.6\frac{1 - 0.364391}{0.600229} = 0.635367$$

For the sum of the values on the right,

$$\sum = 1.000381$$

This is regarded as the solution. Also note that, for a given value of m, this sum increases as $\overline{V}/\overline{L}$ increases.

It follows that

$$L = (L/V)V = (0.4)(12.9312) = 5.1725$$
$$D = V - L = 12.9312 - 5.1725 = 7.7587$$
$$\overline{L} = [1/(\overline{V}/\overline{L})]\overline{V} = [1/0.4](12.9312) = 33.5157$$
$$B = \overline{L} - \overline{V} = 33.5157 - 12.9312 = 20.5845$$

In turn,

$$F = \overline{L} - L = 33.5157 - 5.17225 = 28.3432$$
$$= D + B = 7.7587 + 20.5845 = 28.3432$$

For the overall material balance for component i,

$$28.3432(0.4) \text{ vs. } 7.7587(0.488814) + 20.5845(0.365013)$$
$$11.3373 \text{ vs. } 3.79256 + 7.51361$$
$$11.3373 \text{ vs. } 11.3062$$

For component *j*,

$$28.3432(0.6) \text{ vs. } 7.7587(0.511520) + 20.5845(0.635367)$$
$$17.0059 \text{ vs. } 3.9687 + 13.0787$$
$$17.0059 \text{ vs. } 17.0474$$

The results are considered sufficiently close, therefore, in that the component material balances are largely met. There is an indication that the component material balances are automatically satisfied, as previously pointed out.

The foregoing results apply to $n = 5$ and $m = 5$. For different values of n and m, different results are attained. Predictably, as n and m increase, the sharpness of separation is enhanced. Further, the results are relatively insensitive to variations in L/V and $\overline{V}/\overline{L}$.

These calculations are performed in spreadsheet fashion in Appendix 7 and include the determination for membrane area.

The calculations could be carried one step further, utilizing the relationship between the rectifying section operating lines as developed in Chapter 4; that is, L/D (or L/V) and x_D establish a relationship between $\overline{V}/B$ (or $\overline{L}/\overline{V}$ or $\overline{V}/\overline{L}$) and x_B. This refinement, however, introduces an extra element of trial and error.

Results for Example 7.1

The afore-restated results for Example 4.1 may be compared with the results for Example 7.1 as presented in the spreadsheet calculations in Appendix 7. It may be noted that the entirety of the letter may be made trial and error converging on the ratio B/D.

REFERENCES

1. Hoffman, E. J. *Azeotropic and Extractive Distillation*. New York: Wiley-Interscience, 1964; Huntington, NY: Krieger, 1977.
2. Brown, G. G., A. S. Foust, D. L. Katz, R. Schneidewind, R. R. White, W. P. Wood, G. M. Brown, L. E. Brownell, J. J. Martin, G. B. Williams, J. T. Banchero, and J. L. York. *Unit Operations*. New York: Wiley, 1950.
3. Katz, D. L., D. Cornell, R. Kobayashi, F. H. Poettmann, J. A. Vary, J. R. Elenbaas, and C. F. Weinaug. *Handbook of Natural Gas Engineering*. New York: McGraw-Hill, 1959.
4. Hoffman, E. J. *Phase and Flow Behavior in Petroleum Production*. Laramie, WY: Energon, 1981.

8

Membrane Reactors

With regard to chemically reacting systems, what is called a *membrane reactor* can selectively remove one or another of the products, shifting the conversion to the right. As presented in Chapter 5, a feed-reject crossflow with point permeation is the preferable embodiment, with the effects varying with position (that is, area) along the membrane.

Consider the general case of a chemical reaction bimolecular in both directions, which can be represented by

$$A + B \Leftrightarrow R + S$$

where the capital letters denote the components involved. The reacting system can be homogeneous or catalyzed, that is to say, heterogeneous. Furthermore, for convenience, let the conversion be maintained at a chemical equilibrium, which shifts to the right as one or another of the products is selectively removed. In other words, the rate of removal of the component is slower than the rate of reaction and hence controlling.

Removal of both products from the confines of the reacting system also shifts the conversion to the right, even though the removal of only one component, say, component S, first is considered.

Generally speaking, the reaction equilibrium constant K can be viewed as

$$K = \frac{\gamma_R \gamma_S}{\gamma_A \gamma_B}$$

where γ represents the absolute activities, to be further specified. In terms of the reaction rate constant k_1 for the forward reaction and k_2 for the reverse reaction, the convention is $K = k_1/k_2$. For gaseous systems, partial pressures constitute the convention used, and the reaction equilibrium constant is subscripted K_p. In liquid systems or solutions, concentration is used, and the reaction equilibrium constant can be denoted K_c. Still other conventions may be utilized. With regard to this reaction, there is to be a conversion of X moles per mole of feed introduced. For gaseous systems, therefore, in terms of the number of moles n for each component and a conversion of X moles per mole of feed, it follows that, by definition,

$$\gamma_A = P\frac{n_A}{\sum n} \qquad \gamma_B = P\frac{n_B}{\sum n} \qquad \gamma_R = P\frac{n_R}{\sum n} \qquad \gamma_S = P\frac{n_S}{\sum n}$$

where

$$\sum n = n_A + n_B + n_R + n_S$$

$$n_A = (n_A)_0 - X \qquad n_B = (n_B)_0 - X \qquad n_R = (n_R)_0 + X \qquad n_S = (n_S)_0 + X - Y$$

with the subscript 0 denoting the original moles in existence at the start of the conversion. Furthermore, X can be no greater than either $(n_A)_0$ or $(n_B)_0$.

The original moles of feed is given by the sum

$$\sum n_0 = (n_A)_0 + (n_B)_0 + (n_R)_0 + (n_S)_0$$

(where a likely simplifying assumption is that the number of original moles for each product is zero).

The symbol Y denotes the removal of component S via the accompanying membrane permeation. The basis is the same as for X. Furthermore, Y can be no greater than X.

For a bimolecular reaction, the system pressure cancels out as does the total number of moles $\sum n$, although

$$\sum n = [(n_A)_0 - X] + [(n_B)_0 - X] + [(n_R)_0 + X] + [(n_S)_0 + X - Y]$$

$$= (n_A)_0 + (n_B)_0 + (n_R)_0 + (n_S)_0 - Y$$

Therefore, for the equilibrium constant (or coefficient),

$$K_p = \frac{[(n_R)_0 + X][(n_S)_0 + X - Y]}{[(n_A)_0 - X][(n_B)_0 - X]}$$

Multiplying through and collecting terms,

$$0 = (-K_p + 1)X^2 + \{K_p[(n_A)_0 + (n_B)_0] + [(n_R)_0 + (n_S)_0] - Y\}X$$

$$+ \{-K_p(n_A)_0(n_B)_0 + (n_R)_0(n_S)_0 - (n_R)_0 Y\}$$

$$= aX^2 + bX + c$$

where the coefficients in the quadratic are defined by the substitution. Solving for X from the quadratic,

$$X = \frac{-b \pm \sqrt{b^2 - 4ac}}{2a}$$

and X is therefore a function of Y. The initial equilibrium condition occurs when $Y = 0$.

It may be further stipulated that these quantities are on a unit feedstream molar flow-rate basis.

The gas-phase compositions are of interest and pertain to the feed-reject stream:

$$x_A = \frac{n_A - X}{\sum n} \qquad x_B = \frac{n_B - X}{\sum n} \qquad x_R = \frac{n_R + X - Y}{\sum n} \qquad x_S = \frac{n_S + X - Y}{\sum n}$$

where $\sum n$ also involves Y.

The component selectively removed is designated component S. From the membrane equation for the permeation of component S, it can be written in the appropriate notation that, at a point along the permeation axis, the corresponding permeation molar flux G_S of component S is as follows:

$$G_S = P_S[P_L x_S - P_V y_S]$$

where P_S is the overall membrane permeability coefficient, whose units are defined by the equation. Since only component S is to appear in the permeate, the permeate composition is $y_S = 1$. The reject composition x_S is given by

$$x_S = \frac{(n_S)_0 + (X - Y)}{\sum n} = \frac{(n_S)_0 + (X - Y)}{(n_A)_0 + (n_B)_0 + (n_R)_0 + (n_S)_0 - Y}$$

By virtue of the reaction equilibrium relation, X is a function of Y, and hence x_S becomes a function of Y only. Therefore, on substituting in the flux equation for x_S, the resulting equation is a function of Y, with the initial boundary condition that $Y = 0$ at $A_{membrane} = 0$. The upper limit is as yet arbitrary but will be affixed by the stipulated final conversion for Y, which must be physically real. In turn, the membrane areal requirement $A_{membrane}$ follows. This completes the statement of the problem.

Performing the indicated operation for the equilibrium constant relationship, and assuming for convenience that $(n_R)_0 = 0 = (n_S)_0$, it is obtained symbolically that X is a function $X(Y)$ of Y; that is,

$$X = X(Y)$$

Substituting into the expression for x_S,

$$x_S = \frac{(n_S)_0 + [X(Y) - Y]}{(n_A)_0 + (n_B)_0 + (n_R)_0 + (n_S)_0 - Y}$$

By analogy with Section 5.6 of Chapter 5, for differential permeation with point permeate withdrawal, the equation to be integrated is

$$dV = F\,dY = P_S[P_L x_S - P_V]\,dA_{membrane}$$

or

$$A_{membrane} = \int \frac{F}{P_S[P_L x_S - P_V]}\,dY$$

where F is the total molar feed rate and a constant and, here, $= (n_A)_0 + (n_B)_0$, on a rate basis.

Furthermore, x_S is a function of Y as shown. Although the integration could very possibly be accomplished analytically, for our purposes, a numerical integration suffices, as shown in the accompanying example. Note further that all quantities are consistent and on the same basis.

Again we emphasize that the initial molar quantities can be specified on a feed rate basis; that is, $(n_A)_0$ and $(n_B)_0$ can be specified on the basis of moles per unit time, with the qualification that Y also is in moles per unit time.

Limiting Value for x_S

It may be observed that the integral increases without limit at $P_L x_S \rightarrow P_V$. Accordingly, it is required that

$$P_L x_S > P_V \qquad \text{or} \qquad x_S > \frac{P_V}{P_L} \qquad \text{or} \qquad \frac{P_L}{P_V} > x_S$$

This requirement establishes the pressure ratio across the membrane to reach some arbitrary lower limit or minimal value for x_S.

EXAMPLE 8.1

The following example illustrates the principles involved, with corresponding spreadsheet-type calculations presented in Appendix 8. The bimolecular, bidirectional gas-phase reaction is as follows, as previously indicated:

$$A + B \Leftrightarrow R + S$$

with an assumed reaction equilibrium constant of $K_p = 0.35$. The number of initial moles of each component per mole of feedstream is as follows:

$$(n_A)_0 = 0.4$$
$$(n_B)_0 = 0.6$$
$$(n_R)_0 = 0.0$$
$$(n_S)_0 = 0.0$$

The pointwise membrane permeation coefficient or permeability is 20 in the units of 10^{-9} cm^3(STP)/sec-cm^2-cm Hg/cm, and the membrane thickness is 10 microns or $10(10^{-4})$ cm. The conversion to overall permeability in the appropriate units is

$$20 \frac{76}{22,414} \frac{1}{10(10^{-4})} = 67.8(10^{-9}) \frac{\text{g-moles}}{\text{sec-cm}^2\text{-atm}}$$

The membrane pressures are $P_L = 3$ atm and $P_V = 2$ atm. As per the foregoing derivations, the corresponding spreadsheet calculations and results are displayed in Appendix 8.

Symbols

a	constant or coefficient
A	constant
A	phase A
A	component A
A	membrane area, e.g., in a single-stage separation or in continuous concurrent or countercurrent flow, that is, in differential permeation
A_{membrane}	membrane area in a membrane reactor
A	membrane area for a single cell in the rectifying section of a multistage or cascade operation (all cells in the rectifying section with equal areas); also the area in the rectifying section for differential permeation
A	area of the rectifying section for differential permeation
$\overline{A}$	membrane area for a single cell in the stripping section for a multistage or cascade operation (all cells in the stripping section with equal areas); most generally $\overline{A} = A$
$\overline{A}$	area of the stripping section for differential permeation
A or A_i	absorption factor, $A = L/VK$
b	constant
B	constant
B	component B
B	phase B
B	bottoms or reject product, in moles per unit time
c	constant
c	compressibility of a liquid ($= \beta$, and as distinguished from the compressibility factor z)
$\overline{c}$	mean value for c

c or c_i	concentration
C	constant
d	constant
d	differential operator
D	overhead or permeate product, in moles per unit time
D_i	diffusion coefficient or diffusivity; $D_i = P_i$ in units of (distance)^2/time
F	feed or feedstream designator, moles or molar flow rate; feed to single-stage membrane separator; feedstream to multistage or cascade membrane unit; feedstream to continuous membrane unit
F''	feed or feedstream flux, that is, molar flow rate based on unit area ($F'' = F/A$); and preferably with the area A that of a membrane cell in the rectifying section (which may also equal the area $\overline{A}$ of a cell in the stripping section)
G	mass or molar flux (e.g., for permeate, where $G = V''$)
G_i	mass or molar flux for component i
G_V	mass or molar flux of permeate; $G_V = V''$
$(G_V)_i$	mass or molar flux for component i in the permeate
G_L	mass or molar flux of reject
$(G_L)_i$	mass or molar flux for component i in reject
H	Henry's constant
H	enthalpy or heat function
Hg	mercury symbol
i	component i
I	component i
j	component j
k_g or $(k_g)_i$	gas-film mass transfer coefficient for component i (in partial-pressure units)
k_L or $(k_L)_i$	liquid-film mass transfer coefficient for component i (in partial-pressure units)
k_1	rate constant for forward reaction
k_2	rate constant for reverse reaction
K	reaction equilibrium constant; $K = k_1/k_2$
K	permeability coefficient for fluid flow (in ft^3/hr^2 or ft^3/sec^2)
K or K_i	K-value; for example, $K_i = y_i/x_i$ in terms of phase mole fractions
K_p	reaction equilibrium constant in partial pressure units
K_c	reaction equilibrium constant in concentration units

K'	permeability in darcies (in cg-cm/atm-sec)
K/μ	mobility
K or K_i	permeation K-value or distribution coefficient for component i (including in the "rectifying" section)
$\overline{K}$ or $\overline{K}_i$	permeation K-value or distribution coefficient for component i in "stripping" section, where most generally $\overline{K} = K$
K_i'	permeation K-value or distribution coefficient using recycle
$(K_{A-B})_i$	distribution coefficient for (liquid) phase A mole fraction relative to (liquid) phase B
L	reject or retentate phase, or molar rate, in moles or moles per unit time (e.g., in single-stage separation, or in rectifying section) or in consistent units
$\overline{L}$	reject, or (molar) rate for "stripping" section in moles per unit time, or in consistent units
L''	reject (molar) flux; that is, reject (molar) rate per unit area; $L'' = L/A = G_L$
L^*	dimensionless or reduced reject flux
L'	total reject molar flux using recycle; $L' = L'' + R''$
L_i	molar rate of component i in reject; $L_i = Lx_i$
L_j	molar rate of component j in reject; $L_j = Lx_j$
L_1	reject (molar) rate at point 1
L_2	reject (molar) rate at point 2
m	number of stages or cells in "stripping" section; mth stage or cell
m	constant or coefficient (e.g., in Henry's law)
Δm	membrane thickness
M	substitution quantity; molecular weight
MW	molecular weight
n	exponent (for liquid compressibility behavior)
n	number of stages or cells in "rectifying" section; nth stage or cell
n	total number of moles of reacting components
n_A, and the like	number of moles of component A, and so forth
$(n_A)_0$, and the like	original number of moles of A, and so forth
P or P_i	membrane permeability or mobility to component i in consistent units; $P_i = D_i$ in the units of (distance)^2/time
P or P_i	pointwise membrane permeability; for example, in 10^{-9} cm^3 (at STP)/sec-cm^2-cm Hg/cm, or 10^{-9} cm^2/sec-cm Hg,

	or in g-moles/sec-cm^2-atm/cm, and so on. Defined by its usage.
P or P_i	overall membrane permeability; for example, in 10^{-9} cm^3 (at STP)/sec-cm^2-cm Hg or 10^{-9} cm/sec-cm Hg, or in g-moles/sec-cm^2-atm, and so on. Defined by its usage.
$\overline{P}$ or $\overline{P}_i$	overall membrane permeability (symbol mentioned but not used)
P	pressure
P_0	initial pressure
P_L	high-pressure on reject side of membrane
P_V	low-pressure on permeate side of membrane
$\overline{P}_L$	high-pressure on reject side in stripping section (most usually = P_L)
$\overline{P}_V$	low-pressure on permeate side in stripping section (most usually = P_V)
P_c	critical pressure
P_{pc}	pseudocritical pressure
P_r	reduced pressure
P_0	reference pressure (for liquid compressibility)
R	component R
R	recycle rate, moles per unit time
R''	recycle flux, moles per unit time per unit area
R	pressure ratio
R	gas constant in pressure-volume units
s	distance
S	component S
S or S_i	"stripping" factor, $S = \overline{V}K/\overline{L}$
t	time
t	temperature, centigrade or Celsius scale
T	absolute temperature
T_c	critical temperature
T_{pc}	pseudocritical temperature
T_r	reduced temperature
V	volume (e.g., mass or molar basis)
V_0	initial volume
V	permeate phase, or (molar) rate, in moles or moles per unit time (e.g., in single-stage separation, or in rectifying section)
δV	"drop" of permeate
$\overline{V}$	permeate, or (molar) rate for "stripping" section in moles per unit time, or in consistent units

V''	permeate (molar) flux, that is, permeate molar flow rate per unit area; $V'' = V/A = G_V = G$; permeate (molar) flux in rectifying section
$\overline{V}''$	permeate flux in stripping section
V^*	dimensionless or reduced permeate flux, whereby $V^* = V''/P_i P_V$
V'	total permeate flux using recycle; $V' = V'' + R''$
V_i	molar rate for component i in permeate; $V_i = Vy_i$
V_j	molar rate for component j in permeate; $V_j = Vy_j$
V_1	permeate (molar) rate at point 1
V_2	permeate (molar) rate at point 2
x	linear dimension
x_i	mole fraction of component i (in rectifying section)
$(x_A)_i$	mole fraction of component i in phase A
$(x_B)_i$	mole fraction of component i in phase B
$\overline{x}_i$	mole fraction of component i in the stripping section
$\overline{x}_m$	mole fraction of component in the reject phase leaving the mth stage
x_n	mole fraction of component in the reject phase leaving the nth stage
x_B	mole fraction of component in the bottoms or reject product, $x_B = \overline{x}_1$ in multistage operations
x_D	mole fraction of component i in the overhead or permeate product, $x_D = y_1$ in multistage operations
$(x_F)_i$	mole fraction of component i in the feed
$(x_\Delta)_i$	composition of difference point
$(\overline{x}_F)_i$	constant for component i in single-stage separation: $(\overline{x}_F)_i = b(x_F)_i/(1 - V/F)$
$(\overline{x}_F)_j$	constant for component j in single-stage separation; $(\overline{x}_F)_j = d(x_F)_j/(1 - V/F)$
X	fractional proportion of feedstream that will be reject or retentate
X	degree of conversion (in membrane reactors)
y_i	mole fraction of component i in the permeate phase
$\overline{y}_i$	mole fraction of component i in the stripping section
$\overline{y}_m$	mole fraction of component i in the permeate leaving the mth stage
y_n	mole fraction of component i in the permeate leaving the nth stage
y^*	equilibrium vapor composition
z	compressibility factor (as distinguished from the compressibility c of a liquid)

GREEK LETTERS

α	constant
α	coefficient of expansion for a gas at constant pressure ($\alpha = \alpha^*/V_0$)
α^*	$= \alpha V_0$
α or α_{i-j}	relative permeability, permeability ratio; selectivity; relative volatility
β	constant
β	coefficient of volumetric expansion at constant temperature ($= c$, for a liquid)
$\bar{\beta}$	mean value for β
β	coefficient of pressure expansion for a gas at constant volume ($\beta = \beta^*/P_0$)
β^*	$= \beta P_0$
β or β_{i-j}	selectivity factor
γ_i	absolute activity of component i
γ_i	activity coefficient for component i (in Raoult's law)
δ	constant; variational operator
Δ	difference or difference point
θ	arbitary temperature scale
θ_0	initial value of θ
μ	viscosity
ρ	(mass) density
ρ_0	reference density (e.g., for liquid compressibility)
ρ_m	molar density
φ or φ_{i-j}	relative permeation flux
Φ_i	collection of terms for component i (in rectifying section)
Φ_j	collection of terms for component j (in rectifying section)
$\bar{\Phi}_i$	collection of terms for component i (in stripping section)
$\bar{\Phi}_j$	collection of terms for component j (in stripping section)

SUBSCRIPTS AND DESIGNATORS

A	phase A; component A (in membrane reactors)
B	phase B; component B (in membrane reactors)
F	feed or feedstream
i	component i
j	component j
L	reject
R	component R (in membrane reactors)

S	component S (in membrane reactors)
STP	standard temperature and pressure
V	permeate
0	initial; reference
1	point 1
2	point 2

Appendices

Data and Spreadsheet Calculations

Representative data on membrane permeability and selectivity using selected membranes for various components in gases, liquids, and solutions (or suspensions) are presented in Appendix 1. The spreadsheet layout used illustrates and summarizes the diversity of membrane information. This information, mostly of a random nature, is adapted from the appropriate tabulations in the more readily available literature, and the corresponding references are cited.

Spreadsheet-type calculations may be conveniently used for the presentations and derivations of Chapters 2 through 8; for instance, Lotus 1-2-3, Excel, or whatever. For the purposes here, Excel is utilized. This is contained in Appendices 1 through 8, corresponding to Chapters 1 through 8.

The actual spreadsheet columns are sequenced in alphanumeric, with each appearing as a row designator in the following outlines.

The actual spreadsheet rows correspond to the components involved; that is, each row denotes a particular component. To permit space for the column headings, the identification of the components start at, say, row number 7 on the spreadsheets, for the first (and usually more permeable) component, say, component 1, with the increased numbering continuing downward on the spreadsheets for however many components are to be involved.

The actual spreadsheet calculated results are appended to the corresponding outlines, as pertain to particular chapter examples.

Appendix 1

Representative Membrane Permeabilities and Selectivities

The information provided in the following tables is an assortment of permeability and selectivity values or their alternatives, as per the presentation of the subject in Chapter 1. As the concluding tabulation, in Table A1.23, this information is condensed into spreadsheet representation of selected membrane permeability for various components and membrane materials.

GASES

Table A1.1 Permeability of Solids to Hydrogen at Elevated Temperatures

System	Temp. (°C)	10^{-6} $cm^2/$ sec-atm	10^{-9} $cm^2/$ sec-atm	10^{-9} $cm^2/$ sec-cm Hg
H-Cu	500	3.5	3,500	46.1
	750	8	8,000	105
H-Fe	500	100	100,000	1,316
	600	336	Etc.	4,421
H-Ni	500	3.8		50
	750	31.6		416
H-Pd	500	4,450		58,550
	600	5,750		75,660
H-Pt	600	0.77		0.1
	800	4.8		63.2
H-Zn	300	0.4		5.3
H-rubber	20	0.3		3.9
	100	2.6		34.2
H-SiO$_2$	500		6.2–28	0.08–0.37
	800		35–100	0.46–1.3
H-Pyrex	600		Inappreciable	Inappreciable

Source: Adapted from the *International Critical Tables*, vol. V, pp. 76–77.

Table A1.2   Gas Permeabilities and Selectivity for Different Membranes

Membrane	P_{He}	P_{N_2}	P_{CH_4}	He/N_2	He/CH_4
Silicone	23	15	59	1.5	0.39
Phenyl silicone	15	4.0	20	3.8	0.75
Nitrile silicone	7.9	2.1	10	3.8	0.79
Polycarbonate	6.7	0.46	0.36	15	19
Teflon	6.2	0.25	0.14	25	44
Natural rubber	3.6	1.05	—	3.4	—
Polystyrene	3.5	0.22	0.23	16.0	15
Triethene B	3.4	0.012	0.0084	280	400
Ethyl cellulose	3.1	0.28	0.64	11	4.9
Ethylene/vinyl acetate	2.1	0.28	1.1	7.6	1.9
Viton A	1.7	0.031	0.016	55	110
Polyvinyl chloride	1.4	—	0.2	—	7
Polyethylene	1.0	0.19	—	5.3	—
Polyvinyl fluoride	0.19	0.0019	0.00065	95	280
Mylar (at 25°C)	0.10	0.0006	0.0006	170	170
Saran (at 25°C)	0.066	0.000018	0.000025	370	260

Source: Adapted from R. E. Keating,[1] p. 274. Permeabilities are in units of 10^{-9} cc^3(STP)/cm^2-sec-cm Hg/cm. Temperature is at 30°C except where noted.

Table A1.3   Permeability and Selectivity of Gas in Various Polymer Films

Polymer	P_{He}	P_{H_2}	P_{CO_2}	P_{O_2}	He/CH_4	He/CH_4	CO_2/H_2	O_2/N_2	N_2/CH_4
Rubbery Polymers									
Natural rubber	0.303	0.49	1.34	0.24	1.05	1.63	4.7	2.76	0.30
Silicone	5.61	—	45.53	9.33	0.41	—	3.37	2.12	0.33
Glassy Polymers									
Polysulfone	0.13	0.14	0.056	0.014	49	53	22	5.6	1.0
Polycarbonate	0.14	—	0.065	0.0148	50	—	23.2	5.12	0.93
Cellulose acetate	0.136	0.050	0.055	0.0068	68	48	27.5	3.4	0.73
Polycarbonates									
TCBA-PC or tetrachloro-bis-phenol	—	—	0.026	0.045	—	—	25	6.3	2.2
Poly(pyrrolone)									
6FDA-TADPO* hexafluro-dianhydrides	0.89	—	0.276	0.079	165	—	51.1	6.5	2.4

Table A1.3   (continued)

Polymer	P_{He}	P_{H_2}	P_{CO_2}	P_{O_2}	He/ CH$_4$	He/ CH$_4$	CO$_2$/ H$_2$	O$_2$/ N$_2$	N$_2$/ CH$_4$
Poly(imides)									
6FDA-6fmDa* hexafluro-dianhydride of fluoroenylidene-bis-isopropylaniline	—	—	0.051	0.018	—	—	63.8	6.9	3.26

*The reference supplies the full and exact chemical name.

Source: Adapted from R. R. Zolandz and G. K. Fleming,[2] pp. 45–47. Based on work of R. M. Barrer, "Diffusivities in Glassy Polymers for the Dual Mode Sorption Model," *J. Memb. Sci.* vol. 18 (1984), p. 25, as cited in the chapter references of Zolandz and Fleming. Permeabilities in 10^{-8} cm^3(STP)/cm^2-sec-cm Hg/cm. Temperature is at 25, 30, or 35°C.

Table A1.4   **Relative Gas Permeability for Cellulose Acetate Membranes**

Gas	Relative Permeability
H$_2$O(g) (considered fast)	100
H$_2$	12
He	
H$_2$S	10
CO$_2$	6
O$_2$	1
Ar	
CO	0.3
CH$_4$	0.2
N$_2$	0.18
C$_2$H$_6$ (considered slow)	0.10

Source: Adapted from Gas Research Institute[3] and W. H. Mazur and M. C. Chan.[4]

Table A1.5   **Membrane Permeability to Oxygen**

Polymer	Permeability
Dimethyl silicone	50
Polybutadiene	13
Polyethylene	0.1
Nylon	0.004
Teflon	0.0004

Note: Permeability is in 10^{-9} cm^3(STP)/sec-cm^2-cm Hg/cm.
Source: Adapted from Gas Research Institute.[3]

Table A1.6   Gas Selectivity for Dimethyl Silicone Polymer

Gas	Selectivity ($\alpha = P_i/P_j$)
O_2/N_2	2.0
CO_2/CH_4	3.4
CO_2/H_2	4.9
CO_2/CO	9.0
H_2S/CO	28.0

Source: Adapted from Gas Research Institute.[3]

Table A1.7   Polyamide Membranes for Hydrogen Recovery in Refining and Petrochemicals

Feedsteam	System	% H_2 in Feed	% H_2 in Permeate	% Recovery
Cat reformer	H_2-CH_4	70–80	90–97	75–95+
Cat cracker	H_2-CH_4	15–20	80–90	70–80
Hydroprocessing purge	H_2-CH_4	60–80	85–95	80–95
Adsorber	H_2-CH_4	50–60	80–90	65–85
Ammonia purge	H_2-CH_4	60–70	80–95	85–95
Methanol purge	H_2-CH_4	50–85	90–95	80–95
Benzene recycle	H_2-CH_4	50–60	90–95	85–95
Cyclohexane feed	H_2-CH_4	60–70	90–95	90–95
Synthesis gas	H_2-CO	60–80	90–95	80–95

Adapted from R. R. Zolandz and G. K. Fleming,[2] pp. 81, 84. Based on data from G. L. Poffenbarger and P. Gastinne, with the citation in the chapter references in Poffenbarger and Gastinne.

Table A1.8   Membrane Gas Separations: State of the Technology

Known Separations	To Be Determined
H_2/C_1+	H_2/CO_2
H_2/CO	H_2S/CO_2
He/C_1	NH_3/H_2
$H_2O(g)/C_1+$	NH_3/C_1+
H_2S/C_1+	NH_3/N_2
CO_2/C_1+	SO_2/C_1+
CO_2/N_2	SO_2/CO_2
CO_2/CO	NO_2/C_1+
NO_2/CO	C_1/C_2
NO_2/N_2	N_2/C_1
CO_2/air	Ar/air
	Organic vapors

Note: C_1+ represents methane and heavier hydrocarbons.
Source: Adapted from Gas Research Institute[3] and W. J. Schell.[5]

PERVAPORATION (LIQUID FEED-REJECT/VAPOR PERMEATE)

Table A1.9   Pervaporation Membranes for the Ethanol-Water System

Membrane Material	Ethanol Feed Conc. (Wt %)	Temp. (°C)	Permeate Pressure (kPa)	Selectivity (α)	H_2O Flux $(kg/m^2\text{-}hr)$
Polyvinyl alcohol	92–100	90–100	—	—	0–0.9
	0–100	60	2.0	High	0–5
	60–100	75–100	0.02–5	50–2000	0–2
	0–100	65	0.01	High	0–2.4
Cellulose acetate	0–100	25	0–0.04	5–12	0.1–0.5
Cellulose triacetate	5–95	20	0.01	1–3.6	0.3–1.2
Carboxymethylcellulose	81–95	25	—	2400–5900	0.005–0.1
Polysulfone	15–95	20 to 50	0.01	3–6	—
Acrylic acid-acrylamide	0–90	40	<0.01	1–20	0–10
Polyacrylic acid-polycation	20–100	70	—	<1–2000	0.5–20
Polyvinylfluoride/ acrylic acid	80	70	—	—	1.8
Polyvinylidenefluoride-N-vinylimidazole	0–95	70	3.0	—	0–6
Nafion™	30–98	40	<0.01	Low	<0.5

Note: 1 kPa = 0.001, Mpa = 0.01, bar ~ 0.01, atm ~ 0.15 psi.
Source: Adapted from H. L. Fleming and C. S. Slater,[6] p. 134.

Table A1.10   Representative Water Pervaporation Behavior for Organic/Water Sytstems

Organic Component	H_2O Concentration in Feed (Wt %)	Selectivity (α)	H_2O Flux $(g/m^2\text{-}hr)$
i-Butanol	8.4	1201	1920
THF/benzene	0.255	805	82
Xylene	0.04	5799	25
Methanol	5.1	58	229
Methanol/BTX*	1.1	1823	258
PFP	4.2	22787	1088
Ethanol/benzene	14.1	142	4220
n-Butanol	1.41	929	107
MEK	4.0	3976	907

*BTX refers to benzene, toluene, and xylenes.
Source: Adapted from H. L. Fleming and C. S. Slater,[6] p. 142.

Table A1.11 Organic Pervaporation Behavior of Organic/Water Systems for Selected Membranes

Membrane Material	Organic Feed Conc. (Wt %)	Temp. (°C)	Permeate Pressure (kPa)	Selectivity (α)	Organic Flux (kg/m²-hr)
Polypropylene	Acetone (45)	30	6.5	3	0.1–1.2
Silicone	Butanol (0–8)	30	—	45–65	<0.035
	IPA (27–100)	25	0.33	0.5–12	—
	IPA (9–100)	25	0.67	9–22	0.03–0.11
Polyetheramides	HAc (1.5–9)	50	<0.2	—	0.18–0.28
Polyacrylic acid	HAc (48)	15	—	2–8	0.4–0.55
Silicone	EtAc (0.5–4)	30	−0.2–0.4	High	—
GFT ethanol membrane (PDMS)	Ethanol (87–100)	60	—	150–10,000	0–1.6

Note: 1 kPa = 0.001, Mpa = 0.01, bar ~ 0.01, atm ~ 0.15 psi.
Source: Adapted from H. L. Fleming and C. S. Slater,[6] p. 149.

Table A1.12 Estimated Permeate/Feed Composition Behavior Using Pervaporation to Separate Organic Compounds from Water Using a Silicone Rubber Composite Membrane

Compound	Max. Feed (Wt %)	Max. Permeate (Wt %)	Ratio	Separation Factor
Ethanol*	1.0	6.0	6	6.3
Acetone**	1.0	31.0	31	44.5
Ethyl acetate**	1.0	54.0	54	116.2
1,1,2-Trichloroethane§	0.25	55.0	220	487.7
Chloroform§	0.15	40.0	267	443.8

Note: The feed and reject compositions can be assumed to be equal. The separation factor β_{i-j} can be defined as the ratio of the composition of i to j in the permeate, divided by the ratio of i to j in the reject. In terms of mole or mass fractions for a binary system, where x_i is that for the feed-reject and y_i is that for the permeate, as per Chapter 2, it follows that

$$\text{Separation factor} = \beta_{i-j} \sim \frac{y_i}{1-y_i} \bigg/ \frac{x_i}{1-x_i}$$

where for convenience mass fractions have been used for the last column.

*Hydrophilic.

**Intermediate hydrophobic.

§Hydrophobic

Source: Adapted from J. G. Wigmans, R.W. Baker, and A. L. Aythayde,[7] p. 302.

Table A1.13 Comparative Estimated Separation Factors* for the Pervaporation of Toluene and Trichloroethylene from Water Using Various Rubber Membranes

Membrane	Component	Separation Factor
Fluorocarbon elastomer	Toluene TCE	Small vs. small
Poly-acrylate rubber	Toluene TCE	Small vs. small
Polyurethane	Toluene TCE	Small vs. small
Epichlorohydrin terpolymer	Toluene TCE	~100 vs. ~100
Nitrile butadiene rubber	Toluene TCE	1,000 vs. 500
	Toluene TCE	2,000 vs. 1,000
	Toluene TCE	3,000 vs. 2,000
Polydimethylsiloxane	Toluene TCE	4,000 vs. 3,000
Polynorbornene	Toluene TCE	5,000 vs. 7,000
Nitrile butadiene rubber	Toluene TCE	6,000 vs. 6,000
Polychloroprene	Toluene TCE	18,000 vs. 6,000
Nitrile butadiene rubber	Toluene TCE	30,000 vs. 5,000
Polyoctenamer	Toluene TCE	30,000 vs. 20,000
Ethene-propene-terpolymer	Toluene TCE	50,000 vs. 35,000

*The separation factor for a component i to a component j is defined as the ratio of the concentration of component i to component j in the permeate. This ratio, in turn, is divided by the ratio of the concentration of component i to component j in the reject.

Source: Adapted from J. G. Witmans, R. W. Baker, and A. L. Athayde,[7] p. 303.

LIQUIDS (MISCIBLE)

Table A1.14 Permeate Concentrations in the Separation of Alcoholic and Acidic Aqueous Solutions Using a Hydrophobic Polytetrafluoroethylene (PTFE) Membrane

Solute	Conc. in Feed (Wt %)	Observed Permeate Conc.	Flux $(kg/m^2\text{-}hr)$	Separation Factor
Water	—	—	7.4	—
Methanol	5.0	10.9	9.2	2.3
Ethanol	4.9	12.2	8.8	2.7
2-Propanol	4.5	13.0	10.2	3.2
Formic acid	5.0	2.1	9.2	2.5
Acetic acid	5.0	2.1	8.1	2.5
Propionic acid	5.0	—	9.2	1.2

Source: Adapted from H. E. A. Brüschke and G. F. Tusel,[8] p. 589.

Table A1.15 Permeability of Benzene through Various Treated Cellophane Thicknesses at 20°C

Thickness (mm)	Pressure (kg/cm^2)	Benzene Permeability (ml/pressure-hr-cm^2)
0.0475 (untreated)		
0.058 (treated with 75% EtOH)	36	1.8
0.070 (treated with 75% EtOH)	31	3.4
0.075 (treated with 75% EtOH)	35	3.9
0.082 (treated with dist H$_2$O)	38	13.0

Source: Adapted from R. E. Kesting,[1] p. 86.

Table A1.16 Permeability and Selectivity of Xylenes through Treated Polyethylene

Permeant	Pretreatment	Temp. (°C)	P_i, cm^3 (@STP)/ cm^2-sec-cm Hg × 10^5	Selectivity (p/o)	Selectivity (p/m)	Selectivity (m/o)
Low-Density Polyethylene						
o-Xylene	—	45	1.40			
m-Xylene	—	45	1.55	1.28	1.16	1.10
p-Xylene	—	45	1.79			
o-Xylene	o-Xylene	45	1.89			
m-Xylene	o-Xylene	45	1.87	1.14	1.15	0.99
p-Xylene	o-Xylene	45	2.16			
o-Xylene	m-Xylene	45	1.50			
m-Xylene	m-Xylene	45	1.77	1.39	1.18	1.18
p-Xylene	m-Xylene	45	2.09			
o-Xylene	p-Xylene	45	1.50			
m-Xylene	p-Xylene	45	1.69	1.40	1.24	1.13
p-Xylene	p-Xylene	45	2.10			
High-Density Polyethylene						
o-Xylene	—	30	0.106			
m-Xylene	—	30	0.124	1.62	1.39	1.18
p-Xylene	—	30	0.174			
o-Xylene	p-Xylene	30	1.06			
m-Xylene	p-Xylene	30	1.45	1.62	1.18	1.36
p-Xylene	p-Xylene	30	1.73			
o-Xylene						
m-Xylene						
p-Xylene						

Source: Adapted from R. E. Kesting,[1] p. 87.

SOLUTIONS (SOLUTE SEPARATION)

Table A1.17 Performance of Cellulose Acetate Membranes in Reverse Osmosis

Solute	Test Conditions	Flux × 10^4, cm^3/ cm^2-sec (or gal-ft^2-d)	Percent Rejection
NaCl	50,000 ppm, 8 Mpa	9.17(19.4)	98
Methanol	1.7 Mpa		7
Ethanol	23–138 ppm, 1.7 Mpa		10
Phenol	1.7 Mpa		0
NaCl	5000 ppm, 25°C, 4.1 Mpa, $r = 0\%$	4.8(10.2)	98
Methanol	1000 ppm, 25°C, 4.1 Mpa, $r = 0\%$	4.8(10.2)	<0
Ethanol	1000 ppm, 25°C, 4.1 Mpa, $r = 0\%$		2
Urea	1000 ppm, 25°C, 4.1 Mpa, $r = 0\%$		26
Phenol	1000 ppm, 25°C, 4.1 Mpa, $r = 0\%$		17
NaCl	2000 ppm, 35°C, 2.9 Mpa, $r = 10\%$ pH 5.0–6.0	0.456 liters/sec (or 10,400 gpd)	90
NaCl	2000 ppm, 35°C, 2.9 Mpa, $r = 10\%$ pH 5.0–6.0	0.355 liters/sec (or 8100 gpd)	95
NaCl	2000 ppm, 35°C, 2.9 Mpa, $r = 10\%$ pH 5.0–6.0	0.280 liters/sec (or 6400 gpd)	95
NaCl	2500 ppm, 25°C, 4 Mpa pH 7	19.5(41.3 gal-ft^2-d)	90–92
NaCl	2500 ppm, 25°C, 4 Mpa pH 7	13.9(29.5 gal-ft^2-d)	95–97
NaCl	2500 ppm, 25°C, 4 Mpa pH 7	5.57(11.8)	98–99.5
NaCl	1500 ppm, 25°C, 1.5 Mpa	3.47(7.37)	96
Methanol	1000 ppm, 25°C, 1.5 Mpa		5
Ethanol	1000 ppm, 25°C, 1.5 Mpa		9
Urea	1000 ppm, 25°C, 1.5 Mpa		26
Phenol	1000 ppm, 25°C, 1.5 Mpa		0

Source: Adapted from D. Bhattacharyya, M. E. Williams, R. J. Ray, and S.B. McCray,[9] p. 283.

Table A1.18   Ionic Rejection in Reverse Osmosis

Ion	Feed Conc. (mg/l)	Product Conc. (mg/l)	% Rejection
Calcium	61	0.2	99.6
Sodium	150	3.0	98.0
Potassium	12	0.3	97.4
Bicarbonate	19	0.7	96.2
Sulfate	189	0.4	99.8
Chloride	162	2.9	98.2
Nitrate	97	3.5	96.4
Total dissolved solids	693	11.0	98.4

Source: Adapted from R. G. Sudak,[10] p. 268.

Table A1.19   Cesium Transport through Supported Liquid Membranes

Test No.	Carriers	P_{Cs} (cm-hr^{-1})
1	1,3-Calix[4]-bis-crown-5	9×10^{-2}
2	1,3-Calix[4]-bis-crown-6	1.3
3	1,3-Calix[4]-bis-crown-7	4×10^{-2}
4	1,3-Calix[4]-bis-p-benzo-crown-5	3×10^{-3}
5	1,3-Calix[4]-bis-o-benzo-crown-5	2.8
6	1,3-Calix[4]-bis-napthyl-crown-5	2.7
7	1,3-Calix[4]-bis-diphenyl-crown-5	0.1
8	n-Decyl-benzo-21-crown-5	9×10^{-2}

Notes: Aqueous feed solution: 4M $NaNO_3$ and 1M HNO_3.
Aqueous strip solution: deionized water.
Organic solution: Carrier: 10-2 M in 2-nitrophenyl octyl ether.
Source: Adapted from Z. Asfari et al.,[11] p. 382.

FILTRATION (SUSPENSIONS AND EMULSIONS)

Table A1.20   Membrane Permeability to Water

Type	Material	Initial (Overall) Permeability* (10^{-10} m/s-Pa or m^3/s-m^2-Pa)
Microfiltration	PVDF	472
Ultrafiltration	Polysulfone	11
Ultrafiltration	Polyethersulfone	30
Ultrafiltration	Cellulosic	9
Microfiltration	Polypropylene	140

*During the first few hours the flux dropped sharply to circa 20–30% of the original value, followed by a sort of leveling off with a slight decrease. Backwash was instituted after about 350 hours.
Source: Adapted from P. Aptel,[12] p. 269.

Table A1.21 Water Permeability for Assorted Microfiltration and Ultrafiltration Membranes (at 20°C)

Membrane Material	Pore Size	Membrane Geometry	Permeability (liters/m^2-hr-bar)
Microfiltration			
α-Al_2O_3	0.2 μm	Multichannel	2000
α-Al_2O_3	0.2	Plate	3600
Carbon	0.2	Tubular	1500
SiO_2, Al_2O_3	0.2	Honeycomb	400
SiC	0.2	Multichannel	
Cordierite, mullite	0.5	Spiral-wound hollow	500
SS, Ni, etc.	0.5	Tubular	1300
α-Al_2O_3	0.2	Multichannel	1500
α-Al_2O_3	0.2	Multichannel	2500
Ag	0.2	Tubular plate	9000
SS, Ni, etc.	0.5	Tubular plate	1500
ZrO_2	0.14–0.2	Tubular	600
Ultrafiltration			
γ-Al_2O_3	4 nm	Multichannel	10
	50	Tubular	300
ZrO_2	20	Multichannel	400
	50		800
	100		1500
γ-Al_2O_3	20	Plate	1000
$Zr(OH)_4$-PAA on SS		Tubular	
SiO_2, Al_2O_3	50	Honeycomb	250
ZrO_2 on carbon		Tubular	
SiO_2 (glass)	10	Tubular	
ZrO_2	23	Tubular	70
	83		300
Al_2O_3	50	Tubular	250

Source: Adapted from R. R. Bhave,[13] pp. 103 and 104.

Table A1.22 Effect of Surfactants on the Permeability and Selectivity of Emulsified Toluene Heptane Separations in Transformer Oil

Surfactant	Selectivity Coefficient (maximum)
Rokwinol 60 (polyoxyethylene ester of fatty acids and sorbitan)	2.42
Rokanol K20 (alkyl moiety from coconut oil)	2.95
Sodium oleate	3.65
Potassium palmitate	4.55
Sodium laureate	4.12
Sodium dodecyl sulfate	11.00
Rokafenol N-10 C_9H_{19}-$C_6H_4O(CH_2CH_2O)_{10}H$	3.06
Rokafenol N-8 C_9H_{19}-$C_6H_4O(CH_2CH_2O)_8H$	3.23
Sodium dodecyl benzene	—
Sulfonate	4.60

Notes: Permeate phase: transformer oil.

Feed to Permeate ratio: 1 to 2 by volume.

Surfactant concentration: 0.08 kmoles/m^3.

Source: Adapted from P. Plucinski,[14] p. 478.

SPREADSHEET REPRESENTATION

Table A1.23 Excel Spreadsheet Representation of Selected Membrane Permeabilities

Table	Type of Permeation	Component(s)	Concentration (in wt %)	Membrane	Thickness	Pore Size	Temp.	Feed Pressure	Permeate Pressure
A1.1	gas	H		Cu			500°C		
A1.2	gas	CH_4		Silicone			30°C		
A1.3	gas	H_2		Rubber			25–35°C		
A1.5	gas	O_2		Silicone					
A1.9	pervaporation	$EtOH-H_2O$	0–100% EtOH				25°C		0–04 kPa
A1.10	pervaporation	$1-Butanol-H_2O$	8.4% H_2O						
A1.11	pervaporation	$EtOH-H_2O$	87–100% EtOH	GFT memb			60°C		not given
A1.14	liquids	$EtOH-H_2O$	4.9% EtOH	PTFE	0.075 mm		20°C	35 kg/cm^2	
A1.15	liquids	Benzene		Cellophane					
A1.16	liquids(pervap)	Xylenes		Polyethylene			45°C		
A1.17	reverse osmosis	$NaCl-H_2O$	50,000 ppm	Cellulose acetate				8 Mpa	
A1.19	Cs transport	$NaNO_3-HNO_3$	162 mg/liter	liquid membrane					
A1.20	microfiltration	Water		polypropylene					
A1.20	ultrafiltration	Water		Cellulosic					
A1.21	microfiltration	Water		Al_2O_3		0.2 microm	20°C		
A1.21	ultrafiltration	Water		Al_2O_3		50 nanom	20°C		
2.1, Ex 2.4	liquids (pervap)	$nC7-iC8$	75 vol% $nC7$		1 mil		100°C	15 psig	
2.1, Ex 2.4	liquids (pervap)	$nC7-iC8$	75 vol% $nC7$		1 mil		100°C	115 psig	

Table A1.23 (continued)

Table	10^{-9} cm^3/cm^2-sec-cm Hg/cm (pointwise)	10^{-6} cm^2/sec-atm (pointwise)	ft^2/hr Di in conc units (pointwise)	cm/hr Di in conc units (overall)	cm^3(STP)/cm^2-sec-cm Hg $\times 10^5$ (overall)	ml liq/pressure-hr-cm^2 (overall)	(10^{-10})m^3/s-m^2-Pa (overall)	liters/hr-m^2-bar (overall)	Kg/m^2-hr	Flux $\times 10^4$ in cm^3/cm^2-sec (or gal/ft^2-day)	Flux gal/ft^2-hr $\times 10^3$
A1.1		3.5									
A1.2	59										
A1.3	0.49										
A1.5	50										
A1.9									0.1–0.5 H$_2$O		
A1.10											
A1.11									0–1.6 EtOH		
A1.14									8.8		
A1.15						3.9					
A1.16					1.69						
A1.17										9.17(19.4)	
A1.19				1.3 cm/hr							
A1.20							140				
A1.20							9				
A1.21								2000			
A1.21								250			
2.1, Ex 2.4			0.016(10-4)								140
2.1, Ex 2.4			0.016(10-4)								140

REFERENCES

1. Kesting, R. E. *Synthetic Polymeric Membranes*. New York: McGraw-Hill, 1971.
2. Zoldanz, R. R., and G. K. Fleming. "Gas Permeation: Theory." In *Membrane Handbook*, ed. W. S. Winston Ho and K. K. Sirkar. New York: Van Nostrand Reinhold, 1992.
3. Gas Research Institute. "Proceedings of the First Gas Separations Workshop," Boulder, CO, October 22–23, 1981. Chicago: Gas Research Institute, 1982.
4. Mazur, W. H., and M. C. Chan. "Membranes for Natural Gas Sweetening and CO_2 Enrichment." *Chemical Engineering Progress* 78, no. 10 (1982), p. 38.
5. Schell, W. J. "Membrane Use Technology Growing." *Hydrocarbon Processing* 62, no. 8 (1983), p. 43.
6. Fleming, H. L., and C. S. Slater. "Pervaporation." In *Membrane Handbook*, ed. W. S. Winston Ho and K. K. Sirkar. New York: Van Nostrand Reinhold, 1992.
7. Wigmans, J. G., R. W. Baker, and A. L. Aythayde. "Pervaporation: Removal of Organics from Water and Organic/Organic Separations." In *Membrane Processes in Separation and Purification*, ed. J. G. Crespo and K. W. Böddeker. Dordrecht, the Netherlands: Kluwer Academic Publishers, 1994.
8. Brüschke, H. E. A., and G. F. Tusel. "Economics of Industrial Pervaporation Processes." In *Membranes and Membrane Processes*, ed. E. Drioli and N. Nakagaki. New York: Plenum Press, 1986.
9. Bhattacharyya, D., M. E. Williams, R. J. Ray, and S. B. McCray. "Reverse Osmosis: Design." In *Membrane Handbook*, ed. W. S. Winston Ho and K. K. Sirkar. New York: Van Nostrand Reinhold, 1992.
10. Sudak, R. G. "Reverse Osmosis." In *Handbook of Industrial Membrane Technology*, ed. M. C. Porter. Park Ridge, NJ: Noyes Publications, 1990.
11. Asfari, Z., C. Bressot, J. Vicens, C. Hill, J.-F. Dozol, H. Rouquette, S. Eymard, V. Lamare, and B. Tournois. "Cesium Removal from Nuclear Waste Water by Supported Liquid Membranes Containing Calix-bis-crown Compounds." In *Chemical Separations with Liquid Membranes*, ed. R. A. Bartsch and J. D. Way. ACS Symposium Series 642. Washington, DC: American Chemical Society, 1996.
12. Aptel, P. "Membrane Pressure Driven Processes in Water Treatment." In *Membrane Processes in Separation and Purification*, ed. J. G. Crespo and K. W. Böddeker. Dordrecht, the Netherlands: Kluwer Academic Publishers, 1994.
13. Bhave, R. R. "Permeation and Separation Characteristics of Inorganic Membranes in Liquid Phase Applications." In *Inorganic Membranes: Synthesis, Characteristics, and Applications*, ed. R. R. Bhave. New York: Van Nostrand Reinhold, 1991.
14. Plucinski, P. "The Influence of the Surfactants on the Permeation of Hydrocarbons Through Liquid Membranes." In *Membranes and Membrane Processes*, ed. E. Drioli and N. Nakagaki. New York: Plenum Press, 1986.

Appendix 2

Membrane Permeation Relationships

The units of permeability may appear in several forms, as spelled out in Chapters 1 and 2, albeit the same symbol P_i is used for each set of units. For convenience, the transformation or conversion between the several forms are furnished in spreadsheet notation as follows, with corresponding spreadsheet examples appended for each calculation.

Table A2.1 Excel Spreadsheet Designators and Formulas for Membrane Permeation Relationship Calculations for Example 2.1

Column	Equation or Designator	Spreadsheet Formula
A	Basic P_i (units of 10^{-9} cm^3 per cm^2-sec-cm Hg/cm)	(given)
B	Basic P_i^* $(10^{-3})(76)$, 10^{-6} cm^3 per cm^2-atm/cm or 10^{-6} cm^2 per sec-atm	A10*POWER(10, −3)*76
C	Thickness = Δm (microns or 10^{-4} cm)	(fixed)
D	Overall P_i = Basic P_i^* (10^{-9})* $(76/22,414)$* $[1/\Delta m(10^{-4})]*10^1$ (units of g-moles per cm^2-sec-atm)	A10*POWER(10, −9)* (76/22,414)*(1/C10* POWER(10, −4))*10
E	Overall P_i = D*$(30.48)^2$* $(3600/453.9)$* $(1/14.696)$, lb-moles per ft^2-hr-psi	D10*POWER(30.48,2)* (3600)/(453.9*14.696)
F	Area, cm^2	(set at 1000 cm^2)
G	$P_L - P_V$	(set at 1 atm)
H	Permeation Flux (g-moles/1000 cm^2-sec-1 atm difference, where 929 cm^2 = 1 ft^2)	D10*F10*G10
I	Permeation Flux (lb-moles/ft^2-hr-atm difference)	H10*3600/453.6*(929/ 1000)
J	Diffusivity, cm^2/sec at 1 atm	B10*G10

Table A2.2   Membrane Permeation Relationships—Example 2.1

Basic P_i (given)	Thickness	Overall P_i		Area	$P_L - P_V$	Permeation Flux Rate	Multiplier	Diffusivity
in 10^{-9} cm³ per cm²-sec-cm Hg/ cm or 10^{-6} cm² per sec-cm Hg	Δm in microns or 10^{-4} cm (fixed)	in g-moles per cm²-sec-atm	in lb-moles per ft²-hr-psi	cm² where 929 cm² = one ft² (set)	pressure difference in atm (set)	in g-moles/sec per 1000 cm² per atm diff (per one atm pressure diff)	lb-moles/hr per atm diff	10^{-6} cm²/sec (at one atm)
1.52								
20	10	6.78148E-08	3.40247E-05	1000	1	6.78148E-05	0.0005	1.52

Table A2.3 Excel Spreadsheet Designators and Formulas for Membrane Permeation Relationship Calculations for Example 2.2

Column	Equation or Designator	Spreadsheet Formula
A	Basic P_i (units of 10^{-9} cm^3 per cm^2-sec-cm Hg/cm)	(given)
B	Basic $P_i*(10^{-3})(76)$, 10^{-6} cm^3 per cm^2-atm/cm or 10^{-6} cm^2 per sec-atm	A10*POWER(10, −3)*76
C	Thickness = Δm (microns or 10^{-4} cm)	(fixed)
D	Overall P_i = Basic $P_i*(10^{-9})*(76/22{,}414)*$ $(1/\Delta m(10^{-4})]*10$ (g-moles per cm^2-sec-atm)	A10*POWER(10, −9)* (76/22,414)*(1/C10* POWER(10, −4))*10
E	Overall P_i = $D*(30.48)^2*(3600/453.9)*$ $(1/14.696)$, lb-moles per ft^2-hr-psi	D10*POWER(30.48,2)* (3600)/(453.9*14.696)
F	P_L (10^1 atm)	(set)
G	P_V (10^1 atm)	(set)
H	V/F or V/Feed Note: As per Example 3.1, the symbol F can pertain to the feedstream for the entire stagewise operation	(specified)
I	V (arbitrary units of P_i and P_V or P_L)	(from Example 2.1)
J	Multiplier for Arbitary P_i and P_V: $[(10^{-9})/\Delta m(10^{-4})]*(76/22{,}414)*(10^1)$ will give V (units of g-moles/cm^2-sec)	POWER(10, −9)/(C10* POWER(10, −4))* (76/22,414)*10
K	Permeate Flux: V converted to flux units of g-moles/cm^2-sec	I10*J10*(F10−G10)
L	Membrane Area (cm^2) per g-mole of F per second: F (1 g-mole/sec)*(V/F)/Flux (g-moles/cm^2-sec) gives area in cm^2	H10/K10

Table A2.4 Membrane Permeation Relationships—Example 2.2

Basic P_i (given)		Thickness	Overall P_i		P_L	P_V	V/F	V	Multiplier for Units of P_iP_V	Permeate Flux Rate	Membrane Area per G-mole of Feed per Sec
10^{-9} cm³ per cm²-sec-cm Hg/cm or 10^{-9} cm² per sec-cm Hg	10^{-6} cm³ per cm²-sec-atm/cm or 10^{-6} cm² per sec-atm	Δm in microns or 10^{-4} cm (fixed)	g-moles per cm²-sec-atm	lb-moles per ft²-hr-psi	10^{-1} atm (set)	10^{-1} atm (set)	permeate to cell feed ratio (specified)	molar permeate rate in arbitrary units of P_iP_V (cf. Example 2.1)	$((10^{-9})/[*\Delta m](10^{-4})]]*(76/22414)*10$	g-moles per cm²-sec	(in cm²) Note: 929 cm² = 1 ft²
20	1.52	10	0.00000141	0.00070631	3	2	0.5	12.7056	3.39074E-08	0.00000043	1,160,594.75
10	0.76	10	0.00000070	0.00035316	3	2	0.5	12.7056	3.39074E-08	0.00000043	1,160,594.75

Table A2.5 Excel Spreadsheet Designators and Formulas for Membrane Permeation Relationship Calculations for Example 2.3

Column	Equation or Designator	Spreadsheet Formula
A	Basic P_i (units of 10^{-9} cm^3 per cm^2-sec-cm Hg/cm)	(given)
B	Basic P_i * $(10^{-3})(76)$, 10^{-6} cm^3 per cm^2-sec-atm/cm or 10^{-6} cm^2 per sec-atm	A10* POWER $(10,-3)$*76
C	Thickness = Δm (in microns or 10^{-4} cm)	(fixed)
D	Overall P_i = Basic P_i * $(10^{-9})(76/22,414)$*$[(1/\Delta m(10^{-4})]$*10^1, g-moles per cm^2-sec-atm	A10* POWER $(10, -9)$* $(76/22,414)$
E	Overall P_i = D*$(30.48)^2$*$(3600/453.9)$* $(1/14.696)$, lb-moles per ft^2-hr-psi	D10* POWER $(30.48, 2)$* $(3600)/(453.9*14.696)$
F	Mobility: K/viscosity = Basic P_i $(76)(10^{-9})$, cm^3/cm^2-sec-atm/cm or cm/sec-atm = (centigrams-cm/sec^2-atm) per centigram/ cm-sec = darcies/centipoises	A10*76* POWER $(10, -9)$
G	Gas Viscosity in centipoises (centigrams per cm-sec). One poise = 1 gram per cm-sec, and 100 centipoises = one poise.	(fixed)
H	Permeability in darcies	F10*G10
I	Permeability in millidarcies	H10*1000
J	Permeability in ft^3/hr^2	H10*0.00443

Table A2.6 Membrane Permeation Relationships—Example 2.3

Basic P_i	Thickness	Overall P_i		Mobility K/viscosity	Viscosity	Permeability K		
in 10^{-9} cm³ per cm²-sec-cm Hg/cm or 10^{-9} cm²/(sec-cm Hg)	Δm in microns or 10^{-4} cm (given)	in g-moles per cm²-sec-atm	in lb-moles per ft²-hr-psi	in cm²/sec-atm or darcies per centipoise	in centipoises or in cg per cm-sec (fixed)	in darcies (or cg-cm per atm-sec²)	in millidarcies	in ft³/hr²
20	10	6.78148E-08	3.40247E-05	0.00000152	0.01	1.52E-08	0.0000152	6.7336E-11
1.52								

Table A2.7 Excel Spreadsheet Designators and Formulas for Membrane Permeation Relationship Calculations for Example 2.4

Column	Equation or Designator	Spreadsheet Formula
A	Component Totality	
B	Skip	
C	Sp Gr	(given)
D	P_c (in psia)	(given)
E	MW	(given)
F	Normal boiling point in °F	(given)
G	Normal boiling point in °C	(given)
H	Vapor pressure at 100°F or 37.38°C	(given)
I	Skip	
J	Vol % in Feed-Reject	(specified)
K	Mass fraction in Feed-Reject	J7*C7/(J$7*C$7+J$8*C$8)
L	Mole fraction in Feed-Reject	(J7*C7/E7)/((J7*C$7)/ E$7+(J7*C$8)/E$8)
M	Sp Gr of Feed-Reject	J7*C7/100 M10=SUM(M7:M8)
N	MW of Feed-Reject	E7*L7 N10=SUM(N7:87)
O	Skip	
P	Vol % in Permeate	(specified)
Q	Mass fraction in Permeate	P7*C7/(P$7*C$7+P$8*C$8)
R	Mol fraction in Permeate	(P7*C7/E7)/((P$7*C$7)/E$7+)P$8* C$8)E$8)
S	Sp Gr of Permeate	P7*C7/100 S10=SUM(S7:S8)
T	MW of Permeate	E7*R7 T10=SUM(T7:T8)
U	Skip	
V	Pseudocritical Pressure of Feed-Reject	D7*L7 V10=SUM(V7:V8)
W	Pseudocritical Pressure of Permeate	D7*R7 SUM(W7:W8)
X	Pseudocritical Pressure (average)	(V10+W10)/2
Y	Skip	
Z	Flux in gal/hr-ft^2	(specified)
AA	Flux in lb-moles/hr-ft^2	Z10*(1/7.48)*62.4*S10*(1/T10)
AB	Membrane thickness in mils where 1 mil = 0.001 in. or $0.833(10^{-4})$ ft	(specified)
AC	Skip	

Table A2.7 (continued)

Column	Equation or Designator	Spreadsheet Formula
AD	P_i (calc) for $P_L = 30$ psia	(AA10*R7*AB10)/(30*L7)
AE	P_r for Feed-Reject	30/V10
AF	z	0.17*AE10
AG	z_{av}	AF10/2
AH	D_i in ft^2/hr (calc)	AD10*AG10*10.73*(100+273)*1.8
AI	P_i (calc) for $P_L = 130$ psia	(AA10*R7*AB10)/(130*L7)
AJ	P_r	130/V10
AK	z	0.17*AJ10
AL	z_{av}	AK10/2
AM	D_i in ft^2/hr (calc)	AJ10*AL10*10.73*(100+273)*1.8

Table A2.8   Membrane Permeation Relationships—Example 2.4

						Properties
Component	Sp Gr	P_c (in psia)	MW	nBP°F	nBP°C	vp at 100°F or 37.38°C (in psia)
n-heptane	0.6883	396.8	100.2	209.16	98.42	1.6201
isooctane	0.6962	372.5	114.2	210.63	99.24	1.7089
Totality						

		Feed-Reject (at 100°C)				
Component	vol %	mass frac	mol frac x_i	Sp Gr	MW
n-heptane	50	0.497147	0.529807	0.34415	53.0867
isooctane	50	0.502853	0.470193	0.3481	53.696
Totality				0.69225	106.7827

		Permeate (at 100°C)				
Component	vol %	mass frac	mol frac y_i	Sp Gr	MW
n-heptane	75	0.747854	0.771708	0.516225	77.32519
isooctane	25	0.252146	0.228292	0.17405	26.0709
Totality				0.690275	103.3961

				Flux Relationships		
	Pseudocritical Pressures					Δm
Component	Feed-Reject	Permeate	Average	gal/hr-ft²	lb-moles/hr-ft²	0.833 (10^{-4})
n-heptane	210.227554	306.2139				
isooctane	175.146764	85.03861				
Totality	385.374318	391.2525	388.3134	0.14	0.007797	8.33E-05

		$P_L = 30$ psia			
Component	P_i	P_r (feed-reject)	z	z_{av}	D_i (in ft²/hr)
n-heptane					
isooctane					
Totality	3.15E-08	0.0778464	0.01323	0.006617	1.5E-06

		$P_L = 130$ psia			
Component	P_i	P_r (feed-reject)	z	z_{av}	D_i (in ft²/hr)
n-heptane					
isooctane					
Totality	7.28E-09	0.337334	0.0573	0.028673	1.5E-06

Appendix 3

Single-Stage Membrane Separations

Numbers are used to designate a particular component in column A, and the number 7 may range on up through, say, 16 or however many different components there are to be; that is, i varies from $i = 7$ through $i = 16$, or the number of components involved, as per column A. For a binary mixture, for convenience, i ranges only from $i = 7$ through $i = 8$; that is, $i = 7$ and $j = 8$. Furthermore, the calculation is to be trial-and-error in the variable V such that $\Sigma x_i = 1$ (and $\Sigma y_i = 1$).

The relationships in Table A3.1 apply, as per the accompanying tabulation and spreadsheet calculations, where the asterisk symbol (*) stands for multiplication and the slash symbol (/) for division.

Although the answers appear obvious, in the methodology of Excel, the first-cited equation for L/V is typed into Cell G7 as follows, starting with the equals sign:

$$= 1 - F7$$

In the methodology of Excel, the first-cited equation for K_i is typed into Cell J7 as follows, starting with the equals sign:

$$= C7 * D7 / (I7 + C7 * E7)$$

On pressing the Enter key, the numerical value is entered into Cell J7. To repeat the calculation for the other cells in Column J, press the right-side of the mouse, and on the mini-screen that appears, enter the Copy command. Then highlight or select the additional cells in Column J (using the left-side of the mouse). Again, press the right-side of the mouse, and enter the Paste command that appears on the mini-screen. This will enter the numerical values in the remaining cells in Column J. The same procedure is followed for calculating the x_i in Column K and the y_i in Column L and so forth.

Table A3.1   Excel Spreadsheet Designators and Formulas for Single-Stage Membrane Separation Calculations

Column	Equation or Designator	Spreadsheet Formula
A	Component	
B	$(x_F)_i$	(given)
C	P_i	(given)
D	P_L	(fixed)
E	P_V	(fixed)
F	V/F	(specified)
G	$L/F = 1 - V/F$	G7=1–F7
H (skip column)		
I	V''	(trial and error)
J	$K_i = P_i P_L/(V + P_i P_V)$	J7=C7*D7/(J7+C7*E7)
K	$x_i = (x_F)_i/(V/Fk_i + L/F)$	K7=B7/(F7*J7+G7)
Sum=1	$\Sigma x_i = x_i + x_j$	=K7+K8
L	$y_i = K_i x_i$	L7=J7*K7
Sum (=1)	$\Sigma y_i = $ L7+L6	
M	Skip	
N	Skip	
O7	Δm, microns	(given)
P7	(10^{-9}) (76/22,414) * Q7 * (10^{-4}) * 10	3.39074(10^{-8})
Q	A, cm^2 per g-mole/sec	=F7(1/[(I7)(3.39074(10^{-8})]
R	A, ft^2 per g-mole/sec	= Q7/(929)
S	A, ft^2 per g-mole per hr	=Q7/(3600*929)

To obtain the sum Σx_i for Column K, select the cell in which the sum is to appear; in this particular case, K16. Into this cell type the command

$$=sum(K7:K16)$$

starting with the equals sign. Then, hit Enter. This will give the summation for Cells K7 through K14, or however many cells are to be designated and utilized; similarly for Σy_i in Column L and so on.

ITERATION

It is assumed that the software for Solver has been installed in Microsoft Word initially or as an add-in. Accordingly, using the top toolbar under Tools, select Solver to obtain the pull-down menu.

The Solver pull-down menu has three main categories for the purposes here. The first is called Set Target Cell; here, we enter \$K\$16, signifying Cell K16. Furthermore, the target cell is assigned a value of unity via the subcategory Equal to: Value of 1. This signifies that the value to be targeted is $\Sigma x_i = 1$. (Note that the insertion of the \$ sign is conducted automatically.)

The next principal category is titled By Changing Cells, here we enter \$I\$7, signifying Cell I7. (Note that Cell I6 has already been specified as equal to Cell I7.)

The third category is titled Subject to the Constraints. Under Add there are three entries to complete or designate:

Cell Reference	>=	Constraint
\$I\$7		12

Setting the "greater than or equal" sign has several built-in designations, including the "less than or equal" sign. Setting the constraint may require several trials.

Solver has a secondary pull-down menu, Options, where, according to the Help assistance command, the default settings are noted usually to suffice.

The task is set in motion and completed by hitting Solve. If convergence is not attained, as may be signified in the solution box, then try, try again using different constraints.

The spreadsheet results are shown in Table E.2 for V and the x_i and y_i. The hand-calculated results of Example 3.1 compare favorably. The hand-calculated value for V was 12.8874, provided in Table 3.3, whereas the spreadsheet-calculated value is 12.88487.

For calculational purposes, the determination of the membrane area is based on a membrane thickness of 10 microns, an assumed value of $V'' \sim 12.7$, and a value of $V/F = 0.5$, as in Example 3.1. The nominal membrane pressures of 3 and 2 are replaced by 30 and 20 atmospheres, introducing a factor of 10/1 for the conversion of units.

Other values, of course, may be used, so that the calculation for membrane area can be considered generalized.

Table A3.2 Single-Stage Membrane Separations

Component Number i or j	$(x_F)_i$	P_i to P_j	P_L (fixed)	P_V (fixed)	$V/F < 1$ (specified)	$L/F = 1 - V/F$
1	0.4	20	3	2	0.1	0.9
2	0.6	10	3	2	0.1	0.9
3						
4						
5						
6						
7						
8						
9						
10						
Sum x_i						

Component Number i or j	V'' (trial)	$K_i = P_i P_L / (V + P_i P_V)$	$x_i = (x_F)_i/(V/F) $ $x_i + L/F$	$y_i = K_i x_i$
1	12.88487	1.1345401	0.39469	0.447791
2	12.88487	0.9122737	0.60531	0.552209
3				
4				
5				
6				
7				
8				
9				
10				
Sum x_i			1	1

	Membrane Area per G-mole/sec of Feed for $V/F = 0.5$ and $V'' \sim 12.7$			
Component Number i or j	Δm in microns	Conversion Factor (10^{-9}) * $(76/22414)$ * $[1/10(10^{-4})]$ *(10)	$A = (V/F)$ * $\{(1/[(V'') $ * $3.39074(10^{-8})]\}$ in cm^2	A in ft^2	A in ft^2 per g-mole of feed per hr
1	10	3.39074E-08	1.16E+06	1,250	0.35
2					
3					
4					
5					
6					
7					
8					
9					
10					
Sum x_i					

Appendix 4

Multistage Membrane Separations

The calculations here are confined to two components, the key components, designated i and j (with the components in general designated i).

The feed composition, membrane permeability, and operating pressure levels are assigned as before, and a trial-and-error bubble-point type calculation performed for $V/F = 0$. This results in a value for V''', which is not directly required for determining the degree of separation, other than being incorporated into the K-values but would be necessary in further determining membrane areal requirements. Thus, the bubble-point type calculation simultaneously establishes the values for K_i and K_j, which are used in the separation calculations.

The number of stages n is in turn to be assigned for the rectifying section and the number of stages m for the stripping section (the feed-stream location is $n + 1 = m + 1$).

RECTIFYING SECTION

The trial-and-error calculation for the rectifying section is made dependent on L/V, for both components i and j, in terms of the absorbing factors A_i and A_j. The calculations may be represented as follows, more or less in spreadsheet notation, in terms of component i only:

$$A_i = (x_F)_i \, (1/K_i)$$

$$(\text{Difference})i = [1 - (A_i)^{n+1}] - [1 - (A_i)^n](L/V)$$

Now distinguishing between components i and j:

$$(x_D)_i = K_i \, (x_D)_i \, (1 - A_i)/(\text{Difference})i$$

$$(x_D)_j = K_j \, (x_D)_j \, (1 - A_j)/(\text{Difference})j$$

It is required, as a check, that

$$(x_D)_i + (x_D)_j = 1$$

If the degree of convergence to unity is not satisfactory, then the trial value for L/D or L/V can be adjusted; in fact, the determination can be made computer trial and error.

STRIPPING SECTION

The corresponding calculations are performed for the stripping section in terms of the stripping factors S_i and S_j, which depend on the value for $\overline{V}/B$ or $\overline{L}/\overline{V}$ (or $\overline{V}/\overline{L}$), which is assumed, as well as the values for the K_i, as determined from the bubble-point type calculation on the feedstream composition.

OVERALL MATERIAL BALANCE

Overall, the stream material balance is

$$F = D + B \qquad \text{or} \qquad F'' = D'' + B''$$

And, in spreadsheet notation, for component i or j,

$$F(x_F)_i = D(x_D)_i + B(x_B)_i \qquad \text{or} \qquad F''(x_F)_i = D''(x_D)_i + B''(x_B)_i$$

It may be added that these overall material balances are satisfied automatically by the process of starting at the feed location, where the feedstream compositions are utilized for initiating the calculations in both the rectifying and stripping sections. That is, for convenience in the calculations, the composition of the reject stream leaving the cell at the feed location ($m + 1 = n + 1$) is made identical to the feedstream composition, and the flow rate of the reject in the stripping section is made equal to the flow rate of the reject stream in the rectifying section plus the feedstream flow rate. In other words, in spreadsheet notation,

$$(\overline{x}_{m+1})_i = (\overline{x}_{n+1})_i = (x_F)_i$$
$$\overline{L} = L + F \qquad \text{or} \qquad \overline{L}'' = L'' + F''$$

Other entities, in general, follow the scenario of Example 4.1. Additionally, the degree of separation is expressed in various ways. The overall material balance is checked by

$$\frac{(x_D)_i - (x_F)_i}{(x_F)_i - (x_B)_i} = \frac{B}{D}$$

where spreadsheet notation is used.

MEMBRANE AREA

The membrane area for each cell is the same and, for the purposes here, is evaluated by determining the conversion factor as used in Example 3.1, for the same membrane properties and cell reject and permeate pressures. The conversion factor (divisor) calculates out to $9.26152(10^{-6})$. The conversion is essentially

$$A \text{ (per cell)} = (V/F)(1/V'')/[9.26152(10^{-6})]$$

where $V/F = V''/F''$.

In determining membrane areas, it may be again noted that the numerical values of the permeabilities P_i and P_j are in units of 10^{-9} cm^3/cm^2-sec-cm Hg/cm (where the gaseous volume in cm^3 is at standard conditions). The numerical values of the reject and permeate pressures P_V and P_L are in units of 10^1 atmospheres. And the numerical value of the membrane thickness is in microns or 10^{-4} cm. These units are incorporated into the overall conversion factor.

The foregoing membrane specifications are, of course, to an extent arbitrary and subject to whether the membrane assembly can be made functional for the assigned operating conditions.

However, the spreadsheet calculation sequence itself is designed to allow utilization of other membrane specifications and assigned operating conditions. Moreover, the number of stages in the rectifying and stripping sections can be changed to adjust the degree of separation, as can the recycle or reflux ratios.

In the limit, on the one hand, there would be a condition of minimum reflux or recycle, giving infinite stages, and on the other, total reflux or recycle, whereby no finite product streams are obtained.

As a matter of course, the degree of separation is expressed in a number of different ways and not all that sharp.

FINAL CHECK

A final check can be made in terms of $(x_B)_i$, using the equation

$$x_B = \frac{(\overline{L}/\overline{V} - L/V)x_F + (L/V - 1)x_D}{\overline{L}/\overline{V} - 1}$$

which also automatically satisfies the overall material balance. The values for the feedstream composition $(x_F)_i$ are, of course, initially given, and the values of $(x_D)_i$ are calculated for the rectifying section using an assumed L/D or L/V for a specified number of stages. If the answer is too far afield from the previously calculated values for $(x_B)_i$, then a new value for $\overline{V}/B$ or $\overline{L}/\overline{V}$ can be tried, where

$$\overline{L}/\overline{V} = 1 + B/\overline{V} = 1 + \frac{1}{\overline{V}/B} \quad \text{or} \quad \overline{V}/\overline{L} = \frac{1}{1 + \dfrac{1}{\overline{V}/B}}$$

In fact, the calculation can be made by trial and error in $\overline{V}/B$ or $\overline{L}/\overline{V}$ so that, say, the calculated values for $(x_B)_i$ satisfy the summation requirement that

$$\sum (x_B)_i = 1$$

In Example 4.1, for both the hand calculations and the spreadsheet calculations, the value of $\overline{V}/B$ or $\overline{L}/\overline{V}$ used arrives at a result close enough for most purposes.

Table A4.1 Excel Spreadsheet Designators and Formulas for Multistage Membrane Separations Calculations

Column	Equation or Designator	Spreadsheet Formula
A	Component	
B	$(x_F)_i$	(given)
C	P_i	(given)
D	P_L	(given)
E	P_V	(given)
F	L/D	(assumed)
G	$L/V = 1/[(1/L/D) + 1]$	1((1/F7)+1)
H	$(x_{n+1})_i = (x_F)_i$	B7
I	$(\overline{x}_{m+1})_i = (x_F)_i$	B7
J	Skip	
K	V''	(trial and error)
L	$K_i = P_i P_L/[V + P_i P_V]$	C7*D7/(K7+C7*E7)
M	$\Sigma(y_{n+1})_i = \Sigma K_i(x_{n+1})_i = 1$	L7*H7
N	$(x_{n+1})_i = (y_{n+1})_i/K_i$	M7/L7
O	Skip	
P	$L'' = V'' * (L/V)$	G7*K7
Q	$D'' = L''/(L/D)$	P7/F7
R	Skip	
S	$\overline{V}'' = V''$	K7
T	$\overline{V}/B$	(assumed)
U	$\overline{V}/\overline{L}$	1/(1+(1/T7))
V	$\overline{L}'' = \overline{V}''(\overline{V}/\overline{L})$	S7/U7

Table A4.1 (continued)

Column	Equation or Designator	Spreadsheet Formula
W	$B'' = \overline{L}'' - \overline{V}''$	V7–S7
X	$F'' = \overline{L}'' - L''$	V7–P7
Y	Skip	
Z	$A_i = (L/V)(1/K_i)$	G7*(1/L7)
AA	$1 - A_i$	1–Z7
AB	N	(specified)
AC	M	(specified)
AD	$1 - (A_i)^{n+1}$	1–POWER(Z7,(AB7+1))
AE	$[1 - (A_i)^n](L/V)$	(1–POWER(Z7,AB7)*G7
AF	Difference	AD7–AE7
AG	$(x_D)_i = (y_{n+1})_i(1 - A_i)/\text{Difference}$	M7*(1–Z7)/AF7
AH	Normalized	AG7/SUM(AG7;AG8)
AI	Skip	
AJ	$S_i = (\overline{V}/\overline{L})(K_i)$	U7*L7
AK	$1 - (S_i)^{m+1}$	1–POWER(AJ7,(AC7+1))
AL	$[1 - (S_i)^m](\overline{V}/\overline{L})$	[1–POWER(AJ7,AC7)]*V7
AM	Difference	AK7–AL7
AN	$(x_B)_I = (\overline{x}_{m+1})_i * (1 - S_i)/\text{Difference}$	N7*(1–AJ7)/AM7
AO	Normalized	AN7/SUM(AN7:AN8)
AP	Skip	
AQ	$(x_D)_i/(x_{Fi})$	AH7/B7
AR	$(x_B)_i/(x_{Fi})$	AN7/B7
AS	$(x_D)_i/(x_{Bi})$	AH7/AO7
AT	$(x_B)_i/x_{Di})$	1/AS7
AU	Skip	
AV	$[(x_D)_I - (x_F)_i]/[(x_F)_I - (x_B)_i] = B/D$	(AH7–B7)/(B7–AO7)
AW	B/D	W7/Q7
AX	Skip	
AY	Membrane Thickness: Δm	
AZ	Conversion Factor = (10^{-9}) * $(76/22,414) * [1/\Delta m(10^{-4})] * (10^1)$, g-moles/cm²-sec	POWER(10,–9)* (76/22,414)* (1/AY7(POWER(10,–4))*10
BA	Area per Cell per g-mole of feedstream per sec (cm²)	(K7/X7)*(1/K7)/AY7
BB	Skip	
BC	$[(\overline{L}/\overline{V}) - (L/V - 1)] * (x_{Fi})$	((1/U7)–G7)*B7+(G7–1)*AG7
BD	$(\overline{L}/\overline{V}) - 1$	(1/U7)–1
BE	Check: $(x_B)_i$ $\Sigma(x_B)_i$	BC7/BD7 SUM(BE7:BE8)
BF	Ratio: $(x_B)_i$ Calc/Check	AN7/BE7

Table A4.2 Multistage Membrane Separations

Component Number (i or j)	$(x_F)_i$ (given)	P_i (given)	P_L (fixed)	P_V (fixed)	L/D (assumed)	$L/V = 1/$ $[1/(L/D) + 1]$	$(x_{n+1})_i =$ $(x_F)_i$	$(\bar{x}_{n+1})_i =$ $(x_F)_i$
1 (or i)	0.4	20	3	2	0.666667	0.4	0.4	0.4
2 (or j)	0.6	10	3	2	0.666667	0.4	0.6	0.6
Sum								

Component Number (i or j)	V''' (trial)	$K_i = P_i P_L/$ $[V + P_i P_V]$	$y_i = K_i^*(x_{n+1})_i$ $= (y_{n+1})_i$	$x_i = y_i/K_i =$ $(\bar{x}_{m+1})_i$
1 (or i)	12.93171	1.133536	0.45341438	0.4
2 (or j)	12.93171	0.910976	0.546585591	0.6
Sum			0.999999971	1

Component Number (i or j)	$L'' = (L/V)V''$	$D'' = L''/(L/D)$
1 (or i)	5.172685	7.759028
2 (or j)	5.172685	7.759028
Sum		

Component Number (i or j)	$\bar{V}'' = V''$	$\bar{V}/B$ (assumed)	$\bar{V}/\bar{L} =$ $1/[1 + 1/$ $(\bar{V}/B)]$	$\bar{L}'' = \bar{V}''/$ $(\bar{V}/\bar{L})$	$B'' = \bar{L}'' - \bar{V}''$	$F'' = \bar{L}'' - L''$
1 (or i)	12.93171	0.6666667	0.4	32.32928	19.39757	27.156598
2 (or j)	12.93171	0.6666667	0.4	32.32928	19.39757	27.156598
Sum						

Component Number (i or j)	$A_i =$ $(L/V)^*$ $(1/K_i)$	$1 - A_i$	n (specified)	m (specified)	$1 - (A_i)^{n+1}$	$[1 - (A_i)^n]$ (L/V)	Difference	$(x_D)_i = (y_{n+1})_i^*$ $*(1 - A_i)/$ Difference	Normalized
1 (or i)	0.352878	0.647122	5	5	0.9980692	0.39781132	0.600258	0.488813909	0.48865079
2 (or j)	0.43909	0.56091	5	5	0.9928333	0.39347132	0.599362	0.511519907	0.51134921
Sum								1.000333816	1.000000

Component Number (i or j)	$S_i = \bar{V}^*$ $K_i/\bar{L}$	$1 - (S_i)^{m+1}$	$[1 - (S_i)^m]$ $(\bar{V}/\bar{L})$	Difference	$(x_B)_i =$ $(\bar{x}_{m+1})_i/$ Difference	Normalized
1 (or i)	0.453414	0.991311	0.39233457	0.598976	0.36501313	0.3648743
2 (or j)	0.36439	0.997659	0.39743023	0.600229	0.63536735	0.6351257
Sum					1.00038048	1.0000

Table A4.2 (continued)

Component Number (i or j)	$(x_D)_i/(x_F)_i$	$(x_B)_i/(x_F)_i$	$(x_{Di})/(x_B)_i$	$(x_B)_i/(x_D)_i$
1 (or i)	1.221627	0.912533	1.33923	0.746697
2 (or j)	0.852249	1.058946	0.805115	1.242059
Sum				

Component Number (i or j)	$(x_{Di} - x_{Fi})/(x_{Fi} - x_{Bi}) = B/D$	B/D
1 (or i)	2.523816	2.5
2 (or j)	2.523816	2.5
Sum		

Component Number (i or j)	Memb. Thickness: Δm in microns or 10^{-4} cm (specified)	Conv. Factor: $(10^{-9})* (76/22414)$ $[1/\Delta m(10^{-4})* 10$ (in g-moles/cm^2-sec)	Area per Cell per g-mole of feedstream per sec (in cm^2 where 929 cm^2 = 1 ft^2
1 (or i)	10	3.39074E-08	1,086,001
2 (or j)	10	3.39074E-08	1,086,001
Sum			

Component Number (i or j)	$(\overline{L}/\overline{V} - L/V)(x_F)_i +$ $(L/V - 1)(x_D)_i$	$\overline{L}/\overline{V} - 1$	Check: $(x_B)_i$	Ratio: $(x_B)_i$ Calc/Check
1 (or i)	0.5467117	1.5	0.364474	1.00147801
2 (or j)	0.9530881	1.5	0.635392	0.99996115
Sum			0.999866	

Appendix 5

Differential Permeation with Point Permeate Withdrawal

The objective is to determine the membrane area required for up to 100% transfer of the feedstream $F = L_1$, starting at its bubble point, where $V = V_1 = 0$.

Table A5.1 Excel Spreadsheet Designators and Formulas for Differential Permeation with Point Permeate Withdrawal Calculations

Column	Equation or Designator	Spreadsheet Formula
A	$(x_F)_i$	(given)
B	$(x_F)_j$	(given)
C	P_i	(given)
D	P_j	(given)
E	P_L	(fixed)
F	P_V	(fixed)
G	V/F (signifies bubble-point condition for $L_1 = F$)	(fixed at zero)
H	$L/F = 1 - V/F$	(will be unity)
I	Skip	
J	V or V'' (for bubble-point calc)	(trial and error)
K	$K_i = P_i P_L /(V + P_i P_V)$	C7*E7/(J7+C7*F7)
L	$K_j = P_j P_L /(V + P_j P_V)$	D7*E7/(J7+D7*F7)
M	$x_i = (x_F)_i /[(V/F)K_i + L/F]$	A7/(G7*K7+H7)
N	$x_j = (x_F)_j /[(V/F)K_j + L/F]$	B7/(G7*L7+H7)
O	$y_i = K_i x_i$	K7*M7
P	$y_j = K_j x_j$	L7*N7
Q	$\Sigma x_i = 1$	SUM(O8:P8)
R	$1/(x_i - y_i)$	1/(M7–O7)
S	$[1/(x_i - y_i)]_{av} \, \Delta x_i$	((R7+R8)/2)*(M8–M7)
T	Partial Sum of S	T7+S8

Table A5.1 (continued)

Column	Equation or Designator	Spreadsheet Formula
U	$L/L_1 = 1/\exp(\text{Partial Sum})$	1/EXP(T8)
V	$P_L x_i$	E7*M7
W	$P_v y_i$	F7*O7
X	$P_i(P_L x_i - P_V y_i)$	C7*(V7–W7)
Y	$P_L x_j$	E7*N7
Z	$P_v y_j$	F7*Pj
AA	$P_j(P_L x_j - P_V y_j)$	D7*(Y7–Z7)
AB	$1/[P_i(P_L x_i - P_V y_i) + P_j(P_L x_j - P_V y_j)]$	1/(X7+AA7)
AC	$\Delta A/\Delta V = AB_{av}$, arbitrary units	(AB7+AB8)/2
AD	$\Delta V = -\Delta L$	U7–U8
AE	ΔA, arbitrary units	AC8*AD8
AF	Cumulative Area, arbitrary units	AE7+AE8 where AE7=0
AG	Skip	
AH	Units of P_i and P_j: 10^{-9} cm^3/cm^2-sec-cm Hg/cm	
AI	Units of P_V and P_L: 10^1 atm	
AJ	Δ, microns or 10^{-4} cm	(specified)
AK	Conversion Factor: (10^{-9}) $(76/22,414)[1/10(10^{-4})](10^1) = 9.6152(10^{-6})$	POWER(10,–9)*(76/22,414)* (1/(10*POWER(10,–4)))*10
AL	Cumulative Area per g-mole of L_1 per sec (cm^2, where 929 cm^2 = 1 ft^2)	AF8/AK8

Table A5.2 Differential Permeation with Point Permeate Withdrawal

$(x_F)_i$ (given)	$(x_F)_i$ (given)	P_i (given)	P_i (given)	P_L (fixed)	P_V (fixed)	V/F <1 (specified)	L/F = 1 − V/F
0.4	0.6	20	10	3	2	0	1
0.39	0.61	20	10	3	2	0	1
0.35	0.65	20	10	3	2	0	1
0.3	0.7	20	10	3	2	0	1
0.2	0.8	20	10	3	2	0	1
0.1	0.9	20	10	3	2	0	1
0.05	0.95	20	10	3	2	0	1
0.01	0.99	20	10	3	2	0	1
0	1	20	10	3	2	0	1

V (trial)	$K_i = P_i P_L/(V + P_i P_V)$	$K_i = P_i P_L/(V + P_i P_V)$	$x_i = (x_F)_i/[(V/F)K_i + L/F]$ (check)	$x_i = (x_F)_i/[(V/F)K_i + L/F]$ (check)	$y_i = K_i x_i$	$y_i = K_i x_i$	SUM x_i
12.93172	1.1335358	0.9109759	0.4	0.6	0.453414	0.546586	1
12.84372	1.1354235	0.9134167	0.39	0.61	0.442815	0.557184	0.999999
12.50004	1.1428563	0.9230758	0.35	0.65	0.4	0.599999	0.999999
12.08875	1.1518802	0.9349071	0.3	0.7	0.345564	0.654435	0.999999
11.32383	1.1690476	0.9577373	0.2	0.8	0.23381	0.76619	0.999999
10.62987	1.1850712	0.9794361	0.1	0.9	0.118507	0.881493	1
10.30735	1.1926686	0.9898588	0.05	0.95	0.059633	0.940366	0.999999
10.06031	1.1985543	0.9979937	0.01	0.99	0.011986	0.988014	0.999999
10.00002	1.1999995	0.9999993	0	1	0	0.999999	0.999999

Table A5.2 (continued)

$1/(x_i - y_i)$	$[1/(x_i - y_i)]_{av}$ Δx_i	Partial Sum of $[1/(x_i - y_i)]_{av} \cdot \Delta x_i$	$L/L_1 = 1Exp$ of Partial Sum	$P_L x_i$	$P_V y_i$	$P_j(P_L x_i - P_V y_i)$	$P_L x_i$
-18.72156		0	1	1.2	0.906829	5.863426436	1.8
-18.93395	0.1882776	0.188277583	0.82838473	1.17	0.88563	5.687392968	1.83
-20.00012	0.7786813	0.966958925	0.380237613	1.05	0.799999	5.000011519	1.95
-21.94712	1.0486808	2.015639719	0.133235143	0.9	0.691128	4.177437118	2.1
-29.57747	2.5762292	4.591868894	0.010133901	0.6	0.467619	2.647619147	2.4
-54.03325	4.1805361	8.772404979	0.00015495	0.3	0.237014	1.259715148	2.7
-103.8052	3.9459615	12.71836649	2.9956E-06	0.15	0.119267	0.614662871	2.85
-503.6406	12.148916	24.86728217	1.5859E-11	0.03	0.023971	0.120578282	2.97
				0	0	0	3

$P_V y_i$	$P_j(P_L x_i - P_V y_i)$	$1/Sum\ of$ $\dfrac{P_j(P_L y_i - P_V y_i)}{P_j(P_L x_i - P_V y_i)}$	$\Delta (Area)$ in arbitrary units/ΔV	$\Delta V = (-)\Delta L = (-)\Delta L/L_1$ where $L_1 = 1$	$\Delta (Area)$	Cumulative Area in arbitrary units
1.093171	7.068289774	0.077329257				0
1.1114368	7.156316239	0.077859128	0.077594162	0.17161527	0.013316348	0.013316348
1.199999	7.500013956	0.079999837	0.078929483	0.448147117	0.03537202	0.048868368
1.30887	7.91130003	0.082721627	0.081360732	0.24700247	0.020096302	0.0678467
1.53238	8.676203127	0.088309404	0.085515515	0.123101242	0.010527066	0.079311736
1.762985	9.370149737	0.094074573	0.091191989	0.009978951	0.00091	0.080221737
1.880732	9.692683633	0.09701818	0.095546377	0.000151955	1.45187E-05	0.080236255
1.976028	9.939724974	0.099400582	0.098209381	2.99558E-06	2.94194E-07	0.080236549
1.999999	10.00001359	0.099999864	0.099700223	1.5859E-11	1.58115E-12	0.080236549

Units of P_i and P_j: 10^{-9} cm³ per cm²-sec-cm Hg/cm	Units of P_V and P_L: 10^{-1} atm	Thickness Δm in microns or 10^{-4} cm	Conv. Factor: $(10^{-9})*(76/22414)*[1/\Delta m\,(10^{-4})]*10$ (in g-moles/cm²-sec)	Cumulative Area per g-mole of L_1 per sec (in cm² where 929 cm² = 1 ft²)
		10	3.38923E-08	0
		10	3.38923E-08	392902.3596
		10	3.38923E-08	1436563.12
		10	3.38923E-08	2029509.789
		10	3.38923E-08	2340113.646
		10	3.38923E-08	2366963.447
		10	3.38923E-08	2367391.826
		10	3.38923E-08	2367400.506
		10	3.38923E-08	2367400.506
				2,367,400.51 (Total)

Appendix 6

Differential Permeation with Permeate Flow

The objective is to determine the membrane area for concurrent flow where the feedstream $F = L_1$ is at its bubble point; that is, where $V_1 = 0$.

This procedure may be modified to accept a positive value for V_1 other than zero and a composition $(y_1)_i$. This composition would be entered at E7 = O7 and copied for the entire column. Columns K through R would be left blank, although column O would be made equal to column E.

Table A6.1 Excel Spreadsheet Designators and Formulas for Differential Permeation with Permeate Flow Calculations

Column	Equation or Designator	Spreadsheet Formula
A	$L_1 = F$	(fixed)
B	V_1	(fixed)
C	$(x_1)_i = (x_F)_i$	(fixed)
D	$(x_1)_j = (x_F)_j$	(fixed)
E	$(y_1)_i$ via bubble-point determination for L_1	=O5
F	Skip	
G	P_i	(given)
H	P_j	(given)
I	P_L	(fixed)
J	P_V	(fixed)
K	Skip	
L	V''	(trial and error)
M	$K_i = P_i P_L / (V'' + P_i P_V)$	G7*I7/(L7+G7*J7)
N	$K_j = P_j P_L / (V'' + P_j P_V)$	H7*I7/(L7+H7*J7)
O	$y_i = K_i x_i$, where it also follows that $y_i = (y_1)_i$	M7*C7

289

Table A6.1 (continued)

Column	Equation or Designator	Spreadsheet Formula
P	$y_i = K_i x_i$, where it also follows that $y_i = (y_1)_i$	N7*D7
Q	$\Sigma y_i = 1$	SUM(O7:P7) is required to equal 1
R	Skip	
S	ΔL	(set)
T	$L = L_1 + \Delta L$ etc., where $V_1 = 0$ at $L_1 = F = 1$	T7+S7 where T7=0
U	L_{av}	T7+S7/2
V	Skip	
W	y_i	=0.5
X	x_i	X7=0.4
		X7=AI8
		X9=X8+AI8
Y	Skip	
Z	$P_L x_i$	I7*X7
AA	$P_V y_i$	J7*W7
AB	$\Phi_I = P_i(P_L x_i - P_V y_i)$	G7*(Z7–AA7)
AC	$P_L x_j$	I7*(1–X7)
AD	$P_v y_j$	J7*(1–W7)
AE	$\Phi_j = P_j(P_L x_j - P_v y_j)$	H7*(AC7–AD7)
AF	$\Phi_i/(\Phi_i + \Phi_j)$	AB7/(AB7+AE7)
AG	$-x_i + \Phi_i/(\Phi_i + \Phi_j)$	–X7+AF7
AH	$(1/L)[-x_i + \Phi_i/(\Phi_i + \Phi_j)]$	AH7=AG7
		AH8=(2/U8)*AG8
AI	$\Delta x_i = (1/L)[-x_i + \Phi_i/(\Phi_i + \Phi_j)]\Delta L$	AH7*S7
AJ	x_i	AJ7=X7
		AJ8=X7+AI7
AK	y_i (from material balance)	AK7=O.5
		AK8=–AJ8*(T8/(1–T8))+ 0.4/(1–T8)
AL	Skip	
AM	ΔA, arbitrary units	(1/(AB7+AE7))*(–S7)
AN	Cumulative Area, arbitrary units	AN7=0
		AN8=AN7+AM8
AO	Skip	
AP	Units of P_i and P_j: 10^{-9} cm^3 per cm^2-sec-cm Hg/cm	
AQ	Units of P_V and P_L: 10^1 atm	
AR	Δ, microns or 10^{-4} cm	(specified)
AS	Conversion Factor: (10^{-9}) $(76/22,414)[1/10(10^{-4})](10^1)$ $= 3.39074(10^{-8})$	POWER(10, –9)*(76/22,414)* (1/(10*POWER(10, –4)))*10
AT	Cumulative Area per g-mole of L_1 per sec (cm^2: 929 cm^2 = 1 ft^2)	AN7/AS7

Table A6.2 Differential Permeation with Permeate Flow

$L_1 = F$ (fixed)	V_1 (fixed)	$(x_1)_i = (x_F)_1$ (fixed)	$(x_1)_i = (x_F)_i$ (fixed)	$(y_1)_i$ via bubble pt of L_1 (O7)
1	0	0.4	0.6	0.453414

P_i (given)	P_i (given)	P_L (fixed)	P_V (fixed)
20	10	3	2
20	10	3	2
20	10	3	2
20	10	3	2
20	10	3	2
20	10	3	2
20	10	3	2
20	10	3	2
20	10	3	2
20	10	3	2
20	10	3	2

Table A6.2 (continued)

V" (trial)	$K_i = \dfrac{P_i P_L}{(V'' + P_i P_V)}$	$K_i = \dfrac{P_i P_L}{(V'' + P_i P_V)}$	$y_i = K_i x_i$ or $y_i = (y_1)_1$	$y_i = K_i x_i$ or $y_i = (y_1)_i$	SUM y_i
12.93172	1.133536	0.910976	0.4534143	0.546586	1

ΔL (set)	$L = L_1 + \Delta L$ where $V_1 = 0$ (=L/F)	LAV	y_i	x_i
-0.1	1	0.95	0.453414	0.4
-0.1	0.9	0.85	0.448073	0.394659
-0.1	0.8	0.75	0.445997	0.388501
-0.1	0.7	0.65	0.4388	0.383371
-0.1	0.6	0.55	0.434572	0.376952
-0.1	0.5	0.45	0.429365	0.370635
-0.1	0.4	0.35	0.424033	0.363951
-0.1	0.3	0.25	0.418456	0.356937
-0.1	0.2	0.15	0.412615	0.349542
-0.1	0.1	0.05	0.406477	0.341709
-0.1	0		0.4	0.333346

$P_L x_i$	$P_V y_i$	$(\Phi)_i = P_i(P_L x_i - P_V y_i)$	$P_L x_i$	$P_V y_i$	$(1/L)*AG$
1.2	0.906829	5.863426436	1.8	1.093171	0.05341451
1.183976	0.896146	5.756590502	1.816024	1.103854	0.06157749
1.165502	0.891994	5.47017585	1.834498	1.108006	0.05129372
1.150114	0.8776	5.450285546	1.849886	1.1224	0.06419568
1.130856	0.869144	5.23422319	1.869144	1.130856	0.06316547
1.111906	0.858729	5.063531243	1.888094	1.141271	0.06684076
1.091854	0.848065	4.875773814	1.908146	1.151935	0.07014646
1.07081	0.836911	4.677965535	1.92919	1.163089	0.07394751
1.048626	0.825229	4.467928781	1.951374	1.174771	0.07832553
1.025128	0.812953	4.243487336	1.974872	1.187047	0.08363067
1.000039	0.8	4.000773132	1.999961	1.2	#DIV/0!

$(\Phi)_j = P_j(P_L x_j - P_V y_j)$	$(\Phi)_j/(\Phi)_i + (\Phi)_j$	$(-)x_i + [(\Phi)_i/(\Phi)_i + (\Phi)_j]$
7.068286782	0.453414512	0.05341451
7.121704749	0.44699942	0.05234087
7.264912075	0.429535774	0.04103497
7.274857227	0.4283084	0.04493697
7.382888405	0.414851145	0.03789928
7.468234379	0.404055693	0.03342038
7.562113093	0.39200982	0.02805858
7.661017233	0.379120842	0.02218425
7.76603561	0.365206947	0.01566511
7.878256332	0.350072354	0.00836307
7.999613434	0.333387021	4.0802E-05

Table A6.2 (continued)

$\Delta x_i = AH*\Delta L$	x_i	y_i balance	Δ (Area) in arbitrary units
-0.00534145	0.4	0.453414339	0.007732927
-0.00615775	0.394658549	0.008073061	0.007765003
-0.00512937	0.388500799	0.445996803	0.007852321
-0.00641957	0.383371427	0.438800003	0.007858458
-0.00631655	0.37695186	0.43457221	0.007925744
-0.0066840	0.370635312	0.429364688	0.007979722
-0.00701465	0.363951236	0.424032509	0.008039951
-0.00739475	0.35693659	0.418455747	0.008104396
-0.00783255	0.34954184	0.41261454	0.008173965
-0.00836307	0.341709286	0.406476746	0.008249638
#DIV/0!	0.333346219	0.4	0.008333065

Cumulative Area in arbitrary units	Units of P_i and P_j are in 10^{-9} cm³ per cm²-sec-cm Hg/cm	Units of P_V and P_L are in 10^{-1} atm	Thickness Δm in microns or 10^{-4} cm	Conv. Factor: $(10^{-9})*(76/22414)*$ $[1/\Delta m(10^{-4})]*$ 10 (in g-moles/cm²-sec)	Cumulative Area per g-mole of L_1 per sec (in cm² where 929 cm² = 1 ft²)
0			10	3.39074E-08	0
0.007765003			10	3.39074E-08	229,006
0.015617324			10	3.39074E-08	460,588
0.023475782			10	3.39074E-08	692,350
0.031401526			10	3.39074E-08	926,097
0.039381248			10	3.39074E-08	1,161,436
0.047421199			10	3.39074E-08	1,398,551
0.055525595			10	3.39074E-08	1,637,567
0.06369956			10	3.39074E-08	1,878,634
0.071949198			10	3.39074E-08	2,121,933
0.080282263			10	3.39074E-08	2,367,693
0.080282263 (Total)			10	3.39074E-08	2,367,693

Appendix 7

Countercurrent Flow with Recycle

The preliminaries more or less follow the spreadsheet calculations as presented in Appendix 4. The same feed composition is used, and the same dimensionless permeability and reject and permeate pressures (plus membrane thickness) are given.

The calculations are therefore the same up to and including the bubble-point type determination for the feedstream F, which also establishes the (same) value for the permeate flux V'' (and the corresponding K-values).

However, from this point on, a degree of separation is assigned in terms first of $(x_D)_i$ and then $(x_B)_j$, along with specifying the external recycle ratios L/D and $\overline{V}/B$, and the necessary membrane areas are determined by analytic integration, for both the rectifying sections and stripping sections.

At the end of the dimensionless area calculation, absolute permeability and pressure values are assigned in terms of a conversion (or divisor factor) for a specified membrane thickness, also as in Appendix 4. The area so calculated in cm^2 is placed on the basis of 1 gram-mole per second of feedstream.

RECTIFYING SECTION

The derived integration in terms of component i is

$$\frac{V}{P_i}\frac{1}{\left(P_L\dfrac{V}{L}-P_V\right)}\ln\left\{\left(P_L\frac{V}{L}-P_V\right)y+P_L\frac{D}{L}x_D\right\}\Bigg|_1^2 = A_2$$

where y varies from its value at the feed location ($y = Kx_F$) to its final value x_D. The area A_2 is designated the area for the rectification section.

Integration is from the feed location toward the more-permeable product end, designated D, with A viewed as positive.

The value of V just used can be the flux V'', and to place the area on the basis of the feedstream rate F, the factor $(F''/V'' = F/V)$ can then be introduced. As in Appendix 4, the conversion factor (divisor) to convert the arbitrary or dimensionless units into the prescribed area units of cm^2 is $9.26152(10^{-6})$.

STRIPPING SECTION

The equation used pertains to component j as follows:

$$\frac{\overline{L}}{P_j} \frac{1}{\left(P_L - P_V \dfrac{\overline{L}}{V}\right)} \ln\left\{\left(P_L - P_V \frac{\overline{L}}{V}\right)\overline{x} + P_V \frac{B}{V}x_B\right\}\Bigg|_1^2 = \overline{A}_2$$

where $\overline{x}$ varies from its value at the feed location $(\overline{x} = x_F)$ to its final value x_B.

The preceding provides a determination (estimation) for the membrane area $\overline{A}_2$ in the stripping section based on component j. Integration is from the feed location "downward" toward the less-permeable product end, designated B, with $\overline{A}$ perceived as positive.

The value of $\overline{L}$ used can be the flux $\overline{L}''$, whereby to place the area on the basis of F, the factor $(F''/\overline{L}'' = F/L)$ can then be introduced. As in Appendix 4, the conversion factor (divisor) to convert the arbitrary or dimensionless units into the prescribed units is $9.26152(10^{-6})$.

COMPARISON WITH RESULTS OF APPENDIX 4

The total membrane area calculates to a value pronouncedly less than for the accumulations as obtained in Appendix 4 for an assigned number of stages and, at that, for a much sharper separation.

It can be speculated that the use of a (tubular) membrane as a countercurrent continuum is inherently a much more efficient separation process, or else considerable error occurs from assuming constant molar flow rates in the continuum, or both.

CLOSURE

It may be added that there is not closure on the overall material balances, since B/D as determined from the stream flow rates or fluxes does not agree with that determined from a mole fraction balance. Of course, part of the discrepancy can be traced to the use only of component i in the rectifying section and only j in the stripping section, requiring that the other be obtained in each case by difference.

Calculations eventually become trial and error in the reflux or recycle ratios assigned and in the product compositions assigned. A check may be made on the ratio B/D as indicated in spreadsheet VF and BG.

Table A7.1 Excel Spreadsheet Designators and Formulas for Countercurrent Flow with Recycle Calculations

Column	Equation or Designator	Spreadsheet Formula
A	Component Number	
B	$(x_F)_i$	(given)
C	P_i	(given)
D	P_L	(given)
E	P_V	(given)
F	Membrane Thickness: Δm or Δm (microns or 10^{-4} cm)	(assumed)
G	Skip	
H	L/D	(assigned)
I	$L/V = 1/[(1/L/D) + 1]$	$1((1/H7)+1)$
J	Skip	
K	V''	(trial and error)
L	$K_i = P_i P_L/[V'' + P_i P_V]$	$C7*D7/(K7+C7*E7)$
M	$\Sigma y_i = \Sigma K_i (x_F)_i = 1$	$L7*B7$
N	$x_i = y_i/K_i$	$M7/L7$
O	Skip	
P	$L'' = V'' * (L/V)$	$K7*G7$
Q	$D'' = L''/(L/D)$	$P7/H7$
R	F'' (see AJ5)	$AJ7$
S	Skip	
T	Difference $= P_L(V/L) - P_V$	$D7*(1/I7)-E7$
U	(V/P_i)/Difference (let $V = V''$)	$(K7/C7)/T7$
V	$y_2 = x_D$	(assigned)
W	Difference $* (y_2 = x_D)$	$T7*V7$
X	Difference $* (y_1 = y_i)$	$T7*M7$
Y	Difference $* (y_1) + P_L(D/L) * x_D$	$W7+D7*(1/H7)*V7$
Z	Difference $* (y_1) + P_L(D/L) * x_D$	$X7+D7*(1/H7)*V7$
AA	Y/Z	$Y7/Z7$
AB	Ln(AA)	$LN(AA7)$
AC	A_2 (in arbitrary units)	$U7*AB7$
AD	Conversion Factor $= (10^{-9}) *$ $(76/22,414)[1/\Delta m(10^{-4})] * (10^1)$, g-moles/cm^2-sec	$=AD7=3.39074(10^{-8})$

Table A7.1 (continued)

Column	Equation or Designator	Spreadsheet Formula
AE	A_2, cm^2 per gm-mole of F per sec (929 cm^2 = 1 ft^2)	(R7/K7)*(AC7/AD7)
AF	Skip	
AG	$\overline{V}/B$	(assigned for AG8)
AH	$\overline{L}/\overline{V} = 1 + 1/(\overline{V}/B)$	1+(1/AG8)
AI	$\overline{L}'' = (\overline{L}/\overline{V})*\overline{V}''$	AH8*K8
AJ	$F'' = \overline{L}'' - L''$	AI8–P8
AK	$B'' = \overline{V}''/(\overline{V}/B)$	K7/AG7
AL	Skip	
AM	$\overline{V}/\overline{L}$	1/AH8
AN	Difference = $P_L - P_V * (\overline{L}/\overline{V})$	D8–E8*(1/AM8)
AO	$\overline{L}''/P_j$	AI8/C8
AP	$(\overline{L}/P_j)$/Difference (let $\overline{L} = \overline{L}''$)	AO8/AN8
AQ	$\overline{X}_2 = x_B$	(assigned)
AR	Difference * $(\overline{x}_2 = x_B)$	AN8*AQ8
AS	Difference * $(\overline{x}_1 = \overline{x}_j)$	AN8*N8
AT	Difference * $(x_B + P_V(B/\overline{V}) * x_B$	AR8+E8*(1/AG8)* AQ8
AU	Difference * $\overline{x}_j + P_V * (B/\overline{V}) * x_B$	AS8+E8*(1/AG8)* AQ8
AV	AT/AU	AT8/AU8
AW	Ln(AV)	LN(AV8)
AX	$\overline{A}_2$ (arbitrary units)	AP8*AW8
AY	Conversion Factor = (10^{-9}) * $(76/22,414)[1/\Delta m(10^{-4})] * (10^1)$	=AY8=AD7= 3.39074(10^{-8})
AZ	A_2, cm^2 per gm-mole of F per sec (929 cm^2 = 1 ft^2)	(AJ8/AI8)*(AX8/AY8)
BA	Skip	
BB	$(x_D)_i/(x_F)_i$	V7/B7
BC	$(x_B)_i/(x_F)_i$	(1–AQ8)/B7
BD	$(x_B)_i/(x_D)_i$	(1–AQ8)/V7
BE	Skip	
BF	$[(x_D)_I - (x_F)_i]/[(x_F)_I - (x_B)_i] = B/D$	(V7–B7)/((B7–(1–Q8))
BG	B/D	AK8/Q7

Table A7.2 Countercurrent Flow with Recycle

Component Number (i or j)	$(x_F)_i$ (given)	P_i in 10^{-9} cm^3/ cm^2-sec-cm Hg/ cm (fixed)	P_L in atm(10) (fixed)	P_V in atm(10) (fixed)	Δm in microns or 10^{-4} cm
1 (or i)	0.4	20	3	2	10
2 (or j)	0.6	10	3	2	10
Sum					

Component Number (i or j)	L/D (assigned)	$L/V = 1/$ $[1/(L/D) + 1]$	$V'' = \bar{V}''$ (trial)	$K_i = P_iP_L/$ $[V'' + P_iP_V]$	$y_i = K_i(x_F)_i$	$x_i = y_i/K_i$
1 (or i)	5	0.8333333	12.93171	1.133536	0.453414	0.4
2 (or j)	5	0.8333333	12.93171	0.910976	0.546586	0.6
Sum					1	1

Component Number (i or j)	$L'' = (L/V)V''$	$D'' = L''/(L/D)$	$F'' = A/5$
1 (or i)	10.77643	2.155285	4.741628
2 (or j)	10.77643	2.155285	4.741628
Sum			

Table A7.2 (continued)

Component Number (i or j)	$diff = P_L(V/L) - P_V$	$(V/P_i)/diff$ where $V = V''$	$y_2 = x_D$ (assumed)	$diff*$ $(y_2 = x_D)$	$diff*$ $(y_1 = y_i)$	$diff*$ $(y_2) + P_L$ $(D/L)x_D$	$diff*(y_i) +$ $P_L(D/L)x_D$	Y/Z	$ln(AA)$	A_i in arbitary units	Conv. Factor: $(10^{-9})*$ $(76/22414)*[1/$ $\Delta m(10^{-4})]*$ 10 in g-moles/ cm^2-sec	A_i in cm^2 where 929 $cm^2 = 1\ ft^2$ $(F = 1\ gm$-mole/sec$)$
1 (or i)	1.6	0.404116	0.9	1.44	0.72546302	1.98	1.265463	1.5646447	0.4476588	0.1809061	3.39074E-08	1,125,205
2 (or j)												
Sum												

Component Number (i or j)	$\overline{V}/B$ (assigned)	$\overline{L}/\overline{V} = 1 + 1/(\overline{V}/B)$	$L'' = (\overline{L}/\overline{V}) * \overline{V}''$	$F'' = \overline{L}'' - L''$	$B'' = \overline{V}''/(\overline{V}/B)$
1 (or i)	5	1.2	15.5180554	4.741628	2.58634257
2 (or j)				4.741628	
Sum					

Component Number (i or j)	$\overline{V}/\overline{L}$	$diff = \dfrac{P_L - P_V*}{(\overline{L}/\overline{V})}$	$\overline{L}''/P_i$
1 (or i)	0.833333	0.6	1.551805543
2 (or j)			
Sum			

Component Number (i or j)	$(\bar{L}/P_j)/diff$	$\bar{x}_2 = x_B$ (assumed)	$diff^*$ $(\bar{x}_2 = x_B)$	$diff^*$ $(\bar{x}_1 = \bar{x}_j)$	$diff^* x_B + P_V^* (B/V)^*$	$diff^* \bar{x}_j + P_V^* (B/V)^* x_B$	AT/AU	ln (AV)	$\bar{A}_2$ in arbitrary units	Conv. Factor: $(10^{-9})^*$ $(76/22414)^*$ $[1/\Delta m$ $(10^{-4})] * 10$ in g-moles/cm²-sec
1 (or i)	2.586342571	0.9	0.54	0.36	0.9	0.72	1.25	0.223143551	0.57125666	3.39074E-08
2 (or j)										
Sum										

Component Number (i or j)	$\bar{A}_2$ in cm² where 929 cm² = 1 ft² (F = 1) g-mole/sec	$(x_D)_j/(x_F)_j$	$(x_B)_j/(x_F)_j$	$(x_B)_j/(x_D)_j$	$[(x_D)_j - (x_F)_j]/[(x_F)_j - (x_B)_j] = B/D$	B/D
1 (or i)	3,589,622	2.25	0.25	0.111111	1.666667	1.2
2 (or j)						
Sum						

Appendix 8

Membrane Reactors

The following presentation pertains to an equilibrium chemical conversion that is bimolecular in both directions

$$A + B \Leftrightarrow R + S$$

but may be adjusted for monomolecular behavior in either direction. The reactants are assumed to be at chemical equilibrium and remain at equilibrium during the membrane permeation of product S, which shifts the equilibrium. (Other components could also be permeated, of course, complicating the presentation. Even more so, the chemical conversion rate could be accommodated as well, further complicating the representation.)

The initial moles of reacting components (both reactants and products) are, by definition, specified on the basis of a mole of feed. That is, the initial moles of each, when added up, equal 1 mole of feed. The degree of conversion is denoted by the symbol X. The accompanying membrane transfer of component S, on the same basis, is denoted by Y.

A succession of arbitrary values are assumed for Y, starting at $Y = 0$, corresponding to the initial equilibrium condition. There is an ensuing relationship for X in terms of Y, as determined from the reaction equilibrium constant. This in turn enables the calculation of the simultaneously existing mole fraction x_S of component S, which is also that of the membrane reject stream.

There is a limiting value for x_S, however, by virtue of the integral that establishes the corresponding membrane area:

$$A_{\text{membrane}} = \int \frac{F}{P_S[P_L x_S - P_V]} dY$$

where $y_S = 1$ for a permeate of component S only. Therefore, it is necessary that

$$P_L x_S > P_V \quad \text{or} \quad \frac{P_L}{P_V} x_S > 1$$

Accordingly, the permeation pressures may be adjusted to accommodate this restriction, which becomes a trade-off between the degree of conversion and the upstream membrane feed-reject pressure P_L, for instance, which is also the pressure of the reacting system. (Note that, for the purposes here, a primitive numerical integration is used.)

Table A8.1 Excel Spreadsheet Designators and Formulas for Membrane Reactor Calculations

Column	Equation or Designator	Spreadsheet Formula
A	$(n_A)_0$	(given)
B	$(n_B)_0$	(given)
C	$(n_R)_0$	(given)
D	$(n_S)_0$	(given)
E	K_p	(given)
F	Skip	
G	Y	(values assumed)
H	$a = 1 - K_p$	1–F7
I	$b = K_p[(n_A)_0 + (n_B)_0] + [(n_R)_0 + (n_S)_0 - Y]$	E7*(A7+B7)+(C7+D7)–G7
J	$c = -K_p(n_A)_0(n_B)_0 + (n_R)_0(n_S)_0 - (n_R)_0 Y$	–E7*A7*B7+C7*D7 –C7*G7
K	$\sqrt{(b^2 - 4ac)}$	SQRT(POWER(I7,2)– 4*H7*J7
L	X	(–I7+K7)/(2*H7)
M	$\Sigma n = (n_A)_0 + (n_B)_0 + (n_R)_0 + (n_S)_0 - Y$	A7+B7+C7+D7–G7
N	$x_S = [(n_S)_0 + X - Y]/\Sigma n$	(D7+L7–G7)/M7
O	Skip	
P	P_L	(set)
Q	P_V	(set)
R	$(P_L/P_V)x_S$	(P7/Q7)*N7
S	Skip	
T	P_S, cm^3(STP)/sec-cm^2-cm of Hg/cm	(specified)
U	$P_S[P_L x_S - P_V]$	T7*(P7*N7–Q7)
V	$1/\{P_S[P_L x_S - P_V]\}$	1/U7
W	Skip	
X	$(1/\{P_L[P_L x_S - P_V]\}_{av} * \Delta Y$	(V8+V7)/2*(G8–G7) where X7 = 0
Y	$A = \Sigma(1/\{P_S[P_L x_S - P_V]\}_{av} * \Delta Y$	(Y7+X8) where Y7 = 0
Z	Conversion of membrane area, cm^2/g-mol of feed/sec to ft^2/lb-mole of feed/hr	(Y7/(929))*((473.6)/(3600))

Table A8.1 (continued)

Column	Equation or Designator	Spreadsheet Formula
AA	Skip	
AB	$x_A = [(n_A)_0 - X]/\Sigma n$	(A7–L7)/M7
AC	$x_B = [(n_B)_0 - X]/\Sigma n$	(B7–L7)/M7
AD	$x_R = [(n_R)_0 - X]/\Sigma n$	(C7+L7)/M7
AE	$x_S = [(n_S)_0 + X - Y]/\Sigma n$ (column N)	(D7+L7–G7)/M7

Table A8.2 Membrane Reactors

$(n_A)_0$	$(n_B)_0$	$(n_R)_0$	$(n_S)_0$	K_p	Y (assumed)	a	b	c
0.4	0.4	0	0	0.35	0.00	0.65	0.28	–0.06
0.4	0.4	0	0	0.35	0.05	0.65	0.23	–0.06
0.4	0.4	0	0	0.35	0.10	0.65	0.18	–0.06
0.4	0.4	0	0	0.35	0.15	0.65	0.13	–0.06
0.4	0.4	0	0	0.35	0.16	0.65	0.12	–0.06
0.4	0.4	0	0	0.35	0.17	0.65	0.11	–0.06
0.4	0.4	0	0	0.35	0.18	0.65	0.10	–0.06
0.4	0.4	0	0	0.35	0.19	0.65	0.09	–0.06
0.4	0.4	0	0	0.35	0.20	0.65	0.08	–0.06

$\frac{SQ}{RT}$	X (calc.)	sum n	x_S (calc.)	P_L(atm)	P_V(atm)	$(P_L/P_V)*$ x_S	P_S in (10^{-9}) g-moles/sec-cm^2-atm
0.47	0.15	0.80	0.186	30	2	2.79	6.78E-08
0.45	0.17	0.75	0.154	30	2	2.32	6.78E-08
0.42	0.19	0.70	0.123	30	2	1.84	6.78E-08
0.40	0.21	0.65	0.092	30	2	1.39	6.78E-08
0.40	0.22	0.64	0.087	30	2	1.30	6.78E-08
0.40	0.22	0.63	0.081	30	2	1.21	6.78E-08
0.39	0.23	0.62	0.075	30	2	1.13	6.78E-08
0.39	0.23	0.61	0.069	30	2	1.04	6.78E-08
0.39	0.24	0.60	0.064	30	2	0.96	6.78E-08

$\frac{P_S*}{[P_L x_S - P_V]}$	Reciprocal	Averaged Reciprocal* ΔY	Integral in cm^2 per g-mol of feed/sec	ft^2 per lb-mol feed/hr
2.42E-07	4.13E+06	0.00E+00	0.00E+00	0.00
1.78E-07	5.60E+06	2.43E+05	2.43E+05	32.99
1.15E-07	8.73E+06	3.58E+05	6.02E+05	81.60
5.24E-08	1.91E+07	6.95E+05	1.30E+06	175.89
4.04E-08	2.47E+07	2.19E+05	1.52E+06	205.60
2.86E-08	3.50E+07	2.99E+05	1.81E+06	246.09

Table A8.2 (continued)

P_S^* $[P_L x_S - P_V]$	Reciprocal	Averaged Reciprocal* ΔY	Integral in cm^2 per g-mol of feed/sec	ft^2 per lb-mol feed/hr
1.7E-08	5.89E+07	4.69E+05	2.28E+06	309.75
5.59E-09	1.79E+08	1.19E+06	3.47E+06	471.01
−5.55E-09	−1.80E+08			

x_A	x_B	x_R	x_S (same as column N)
0.314	0.314	0.186	0.186
0.312	0.312	0.221	0.154
0.306	0.306	0.266	0.123
0.292	0.292	0.323	0.092
0.288	0.288	0.337	0.087
0.284	0.284	0.351	0.081
0.280	0.280	0.365	0.075
0.275	0.275	0.381	0.069

Index